3rd International Conference on Science and Technology Applications in Climate Change (STACLIM 2018)

Fourth Industrial Revolution for Sustainable Environmental Application

IOP Conference Series: Earth and Environmental Science
Volume 228

AA006277

Ayer Keroh, Malaysia
13 – 15 November 2018

ISBN: 978-1-5108-8156-3
ISSN: 1755-1307

Printed from e-media with permission by:

Curran Associates, Inc.
57 Morehouse Lane
Red Hook, NY 12571

Some format issues inherent in the e-media version may also appear in this print version.

This work is licensed under a Creative Commons Attribution 3.0 International Licence.
Licence details: http://creativecommons.org/licenses/by/3.0/.

No changes have been made to the content of these proceedings. There may be changes to pagination and minor adjustments for aesthetics.

Printed by Curran Associates, Inc. (2019)

For permission requests, please contact the Institute of Physics
at the address below.

Institute of Physics
Dirac House, Temple Back
Bristol BS1 6BE UK

Phone: 44 1 17 929 7481
Fax: 44 1 17 920 0979

techtracking@iop.org

Additional copies of this publication are available from:

Curran Associates, Inc.
57 Morehouse Lane
Red Hook, NY 12571 USA
Phone: 845-758-0400
Fax: 845-758-2633
Email: curran@proceedings.com
Web: www.proceedings.com

Table of contents

Volume 228

Proceedings of the 3rd International Conference on Science and Technology Applications in Climate Change

13–15 November 2018, Holiday Inn Hotel, Ayer Keroh, Malaysia

Accepted papers received: 1 July 2019 Published online: 15 February 2019

Preface

Papers

IOP Conf. Series: Earth and Environmental Science **228** (2019) 011001 doi:10.1088/1755-1315/228/1/011001

Preface: Proceedings of the International Conference on Science and Technology Applications in Climate Change (STACLIM 2018)

The 2018 International Conference on Science & Technology Applications in Climate Change (STACLIM 2018) took place in Melaka, Malaysia between November, 14 and November 15, 2018. STACLIM 2018 was organised by the Institute of Climate Change, Universiti Kebangsaan Malaysia and involves the cooperation of Universiti Teknikal Malaysia Melaka (UTeM). The conference was the continuation of the 2014 and 2016 STACLIM conference series. This biennial conference series aims to confront various challenges pertinent to climate change and proffer opportunities to better understand how science and technological applications can play significant roles for drawing the immediate attention of people and providing action plans.

The third series of the conferences, with the theme "Fourth Industrial Revolution (4IR) for Sustainable Environmental Application", continues with the intention of bringing together scientists, engineers, students and researchers from all over the world to present and discuss their research findings, and offer them a platform for enabling their contributions to address global climate change issues. The conference has attracted participants from the United States, United Kingdom, India, China, and local.

This conference is expected to provide an excellent opportunity to foster discussion and collaboration among researchers working in the fields of Environmental Science, Health Sustainability, Mathematics, Sustainable Energy, Economic Sustainability, Socio-Cultural Studies, Social Science, Atmospheric Science, and others. The conference is joined by five keynote speakers who are leading experts from reputable organizations such as Dr. Matthew Ashfold from University of Nottingham Malaysia Campus; Prof. Dr. Gaopeng Lu from Chinese Academy of Sciences, Beijing, China; Assoc. Prof. Dr. Maizatun Mustafa from International Islamic University Malaysia (IIUM); Assoc Prof. Dr. Rawshan Ara Begum from Kumamoto University, Japan and Dr. Alison Copeland from Newcastle University, United Kingdom. The keynote speeches covered various themes and sessions presented at the conference such as Internet of Things for Atmospheric Observation, Space Weather, Policy in Climate Change, Economic Sustainability as well as Environmental Health and Sustainability.

We would like to thank all members that participated in any way in the STACLIM 2018 conference especially to the co-organising university for their support and development of a high-quality conference scientific level and profile, to the members of the Scientific Committee that honoured the conference with their presence and provided a significant contribution to the

Content from this work may be used under the terms of the Creative Commons Attribution 3.0 licence. Any further distribution of this work must maintain attribution to the author(s) and the title of the work, journal citation and DOI.
Published under licence by IOP Publishing Ltd

review of papers. Special thanks should also be given to all members of the Organising Committee for their help, support and spirit participation before, during and after the conference. Lastly, our deepest appreciation also goes to the keynote speakers, authors, and everyone, who contributed directly or indirectly from all over the world, whose support and encouragement is of paramount importance towards the success of this conference. The STACLIM 2018 proceeding is a credit to a large group of people, and everyone should be proud of the outcome.

Peer review statement

All papers published in this volume of *IOP Conference Series: Earth and Environmental Science* have been peer reviewed through processes administered by the proceedings Editors. Reviews were conducted by expert referees to the professional and scientific standards expected of a proceedings journal published by IOP Publishing.

Content from this work may be used under the terms of the Creative Commons Attribution 3.0 licence. Any further distribution of this work must maintain attribution to the author(s) and the title of the work, journal citation and DOI.
Published under licence by IOP Publishing Ltd

IOP Conf. Series: Earth and Environmental Science **228** (2019) 012001 doi:10.1088/1755-1315/228/1/012001

Anthropogenic and biogenic volatile organic compounds and ozone formation potential in ambient air of Kuala Lumpur, Malaysia

H H A Hamid[1,2], M H M Hazman[2], M S M Nadzir[2,3], R Uning[2], M T Latif[2], N Kannan[4]

[1]Institute for Environment and Development (LESTARI), Universiti Kebangsaan Malaysia, 43600 UKM, Bangi, Selangor, Malaysia

[2]School of Environmental and Natural Resource Sciences, Faculty of Science and Technology, Universiti Kebangsaan Malaysia, 43600 UKM, Bangi, Selangor, Malaysia

[3]Centre for Tropical Climate Change System, Institute of Climate Change, Universiti Kebangsaan Malaysia, 43600 Bangi, Selangor, Malaysia, Malaysia

[4]Faculty of Applied Sciences, AIMST University, 08100 Bedong, Kedah, Malaysia

E-mail: haris@ukm.edu.my

Abstract. The measurement of anthropogenic and biogenic volatile organic compounds (VOCs) in urban air is important because of their toxic effects and because they can produce secondary air pollutants (e.g. ozone, particulate matter). Benzene, toluene, ethylbenzene and xylenes (BTEX), representing anthropogenic VOCs (AVOCs), and isoprene, a biogenic VOC (BVOC), were monitored at an urban site in Kuala Lumpur (KL). Active sampling was conducted using sorbent tubes packed with Tenax®TA/1 TD for 5 days (daytime). Analysis of VOCs was performed using Thermal Desorption-Gas Chromatography Mass Spectrometry (TD-GCMS). Average daytime concentrations of VOCs were 9.42 ± 6.15 μg/m^3 for benzene, 35.04 ± 12.60 μg/m3 for toluene, 11.20 ± 4.17 μg/m^3 for ethylbenzene, 19.19 ± 10.76 μg/m^3 for m,p-xylene, 18.40 ± 8.56 μg/m^3 for o-xylene, and 6.70 ± 1.57 μg/m^3 for isoprene. Weekday BTEX levels were found to be higher, especially during the morning rush hour, demonstrating the impact of traffic emissions in KL. BVOCs (isoprene) in urban KL was also higher than the reported isoprene level in Malaysian forest areas but lower than palm plantation areas. Both AVOCs and BVOCs significantly contributed to (218.6 – 733.8 μg/m^3) ozone formation in the KL atmosphere.

1. Introduction

Volatile organic compounds (VOCs) can be emitted from anthropogenic and biogenic sources to the atmosphere. Anthropogenic VOCs (AVOCs) such as benzene, toluene, ethylbenzene, and xylene isomers (BTEX) are dominant in the urban atmosphere due to vehicle, industrial, and biomass emissions [1][2][3]. Meanwhile, biogenic VOCs (BVOC), mainly isoprene, are released by plants and vegetation [4][5]. The measurement of AVOCs and BVOCs is important due to their contribution to surface ozone formation, secondary organic aerosol (SOA), the health risk posed to humans, and their

Content from this work may be used under the terms of the Creative Commons Attribution 3.0 licence. Any further distribution of this work must maintain attribution to the author(s) and the title of the work, journal citation and DOI.
Published under licence by IOP Publishing Ltd

environmental impact [6]. Moreover, BTEX are hazardous air pollutants (HAPs) and benzene is the most toxic and a known carcinogen.

The impacts of isoprene and BTEX on urban air may vary depending on meteorological and regional differences [7][8]. Tropical regions receive high solar radiation throughout the year this may induce photochemical reactions of VOCs and influence VOC levels and atmospheric composition [9]. Isoprene can play a very significant role in ozone formation due to its higher reactivity compared to other BVOCs [10]. Both BVOCs and semivolatile organic compounds (SVOCs) produce SOA which directly and indirectly impacts climate change via solar radiation scattering [11]. Even though BVOCs are known to be emitted at greater levels compared to AVOCs globally, in the case of urban areas, AVOCs may significantly influence the local ambient air [12].

The Kuala Lumpur (KL) urban area, similar to other cities in South East Asia and developing countries, is facing air pollution problems related to heavy traffic [13]. VOCs, including BTEX, are released into the atmosphere and polluted the surrounding areas from incomplete combustion in vehicle engines [14][15]. A Study by Lan & Binh [16] revealed that the benzene level at the roadside in Kuala Lumpur was 48 $\mu g/m^3$. Meanwhile, the outdoor level of benzene in Johor Bahru, another urban city in Malaysia, was measured at 5.1 $\mu g/m^3$ [17]. However, information on BVOCs, i.e. isoprene, in the ambient air of urban KL is still very limited.

This paper aims to evaluate the levels of isoprene and BTEX, representing biogenic and anthropogenic VOCs, in the ambient air of urban KL. Ozone formation potential was also calculated to understand their contribution to surface ozone.

2. Methodology

2.1 Sampling

Sampling was conducted at the Faculty of Health Science, Universiti Kebangsaan Malaysia, Kuala Lumpur (UKM, KL) (N03° 10'5.28", E101° 42'1.56") (Figure 1). The sampling was performed on the 3rd floor (~10 m from the ground) and the building is situated near two major KL roads, Jalan Raja Muda Abdul Aziz and Jalan Pahang. Five days of sampling (30th January 2018 – 3rd February 2018) were undertaken at three daytime intervals; T1 (0800-1000), T2 (1200-1400), T3 (1600-1800). The sampling procedure was based on the active sampling method, using Tenax®TA/1 TD (Markes International, UK) multi-sorbent tubes. A low-flow pump (Personal Air Sampler) PAS-500 (Supelco, US) was set at a flow rate of 60 ml/min and operated for 2 h.

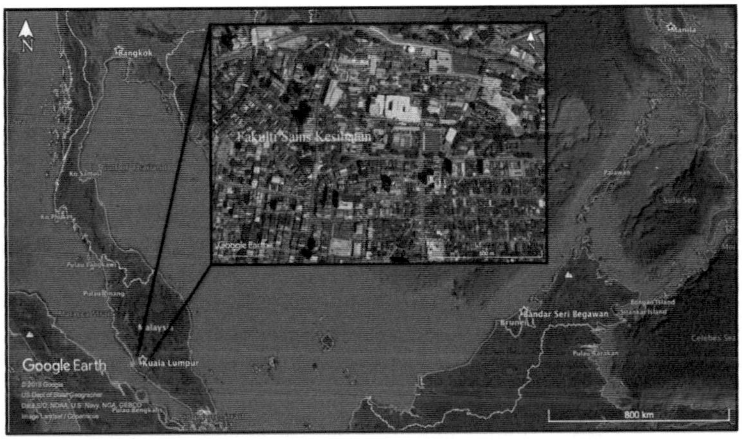

Figure 1. VOC sampling location map in Kuala Lumpur

2.2 *Analysis*

The analysis of the sorbent tubes was performed using thermal desorption (TD) (Markes, International, UK) gas chromatography mass spectrometry (GCMS) (Agilent Technologies, US). The TD-GCMS system was equipped with a DB 624 column (60 m × 0.25 mm × 0.25 μm) (Agilent J&W Scientific, US) for VOC separation. The tubes were thermally desorbed at 280°C for 30min onto the focusing trap at 25°C, and the trap high was performed at 330°C for 3 min to transfer the analytes to the GCMS. The temperature program was set at initial 36°C (4min), with a ramp rate of 5°C/min to 250°C. The measurement of isoprene and BTEX was based on an external calibration. Gas standards of 10 ppm BTEX (MESA, USA) and 100 ppm isoprene (MESA, USA) were diluted with purified nitrogen to several individual BTEX and isoprene concentrations (30-400 ng) in the sorbent tubes for the multiple point calibration. A calibration solution loading rig (CSLR), (Markes, UK) was used as an aid to preparing the standards in the stainless steel sorbent tubes. Calibration linearity results (r^2) for all compounds were good, >0.987. Validation was performed using different standard gases containing 100 ppb of 62 toxic gas VOCs (TO-17 gas standards) (Scotty, US). Recovery for BTEX was found 81-104%.

2.3 *Ozone Formation Potential (OFP)*

OFP can be a reference metric to describe the potential maximum ozone formation in urban areas where VOC-sensitive [18]. The contribution of individual BVOCs and AVOCs to OFP was studied using their Maximum Incremental Reactivity (MIR). According to Duane et al. [19], MIR can be referred to as the maximum amount of ozone produced by the photolysis of specific volatile compounds. OFP can be measured as:

$$\text{OFP } (\mu g/m^3) = \text{concentration } (\mu g/m^3) \times \text{MIR coefficient value}$$

The MIR coefficient values used were determined based on updated MIR values (2009) by the California Air Resources Board (CARB, 2009) [assessed from webpage https://www.arb.ca.gov/regact/2009 on 17082018].

IOP Conf. Series: Earth and Environmental Science **228** (2019) 012001 doi:10.1088/1755-1315/228/1/012001

3. Results & Discussion

3.1 Level of daytime VOCs

Levels for every compound were measured during the sampling period and the results are shown in Table 1. Average concentrations during the daytime sampling period were 6.70 ± 1.57 μg/m^3 for isoprene, 9.42 ± 6.15 μg/m^3 for benzene, 35.04 ± 12.60 μg/m^3 for toluene, 11.20 ± 4.17 μg/m^3 for ethylbenzene, 19.19 ± 10.76 μg/m^3 for m,p-xylene and 18.40 ± 8.56 μg/m^3 for o-xylene. Toluene showed the highest levels measured for all sampling days with a range of 6.32 – 67.08 μg/m^3. The lowest concentrations were measured for benzene during the T2 sampling (1/2/18) and the weekend (3/2/2018). Levels for AVOCs for most of the sampling time followed the pattern toluene > m,p-xylene > o-xylene > benzene/ethylbenzene. The highest BTEX levels occurred during T1. As illustrated in Figure 2, average VOC concentrations during T1 and T3 were high compared to T2 (12-2pm), especially during weekdays. The levels of BTEX were high during T1 and T3 due to the rush hour when traffic on the nearby roads was at its heaviest. As reported by Phuc and Kim Oanh [20], high levels of BTEX related to traffic often occur in SEA cities. The average BTEX level (93.25 μg/m^3) found in KL was similar to those reported in other SEA urban cities: Bangkok, Thailand (107.6 μg/m^3); Ho Chi Minh City, Vietnam (98.2 μg/m^3); but lower than Manila, Philippines (275.1 μg/m^3)[21][22][23].

Daytime isoprene levels were in the range 3.05 – 10.52 μg/m^3 (1.21 – 3.87 ppbv) and the levels were found to be slightly lower during the afternoon T3 sampling. The levels were found to be higher than average isoprene concentrations in a forest area (~1.5ppbv) in Borneo, Malaysia but lower than a palm oil plantation area (~4 ppbv) in Tawau, Sabah [24]. Saito et al. [25] also reported daytime isoprene levels in the range 0.3 – 3.0 ppbv in a reserve forest in Negri Sembilan, Malaysia. The isoprene within urban KL might be released by surrounding forest areas and palm plantations on the outskirts of the KL urban area. During the sampling period, the post-northeast monsoon (Jan-Mar) season occurred and possibly transported isoprene emissions from palm plantations in the states of Pahang and Perak. According to Hellen et al. [12], isoprene also can be emitted by traffic.

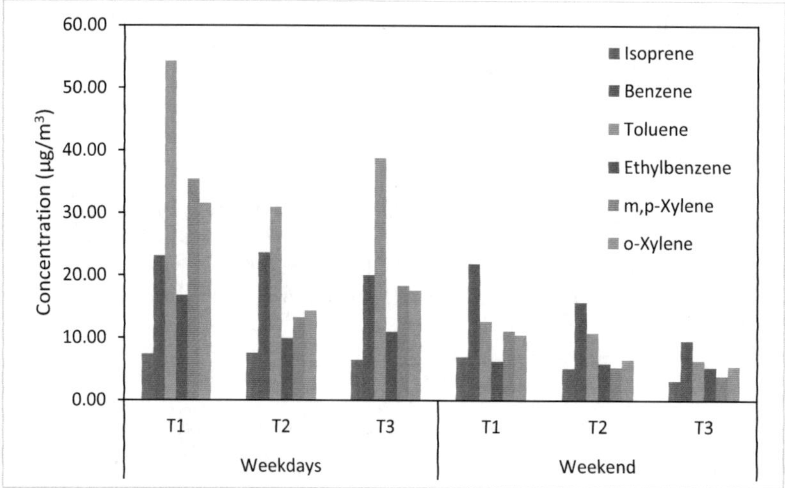

Figure 2. Average concentrations (μg/m^3) of isoprene and BTEX during weekdays and the weekend

3.2 Ozone Formation Potential

The measured levels of VOCs were used to calculate the ozone formation potential (OFP) as shown in Table 2. The range of ∑OFP for the studied VOCs during the sampling period was 218.6 – 733.8 µg/m3. Average OFPs produced by individual compounds were led by m,p-xylene (145.8 µg/m^3) and followed by o-xylene (139.9 µg/m^3), toluene (135.9 µg/m^3), isoprene (71.5 µg/m^3), ethylbenzene (32.8 µg/m^3) and benzene (6.50 µg/m^3). The dominant contributors to OFP in KL were m,p-xylene, o-xylene and toluene and benzene was the lowest due to its low reactivity. The highest OFP occurred on 30/1/2018 and the lowest (218.6 µg/m^3) was measured on the weekend of 3/2/2018. The contribution of isoprene to ∑OFP was calculated as >10% in the KL ambient air and the level increased to 24.5% during the weekend. In this study, AVOCs showed more effects on the local tropospheric ozone in the urban atmosphere. The results for OFP indicated that AVOCs and BVOC have significant impacts on surface ozone in urban KL.

4. Conclusion

Levels of anthropogenic (BTEX) and biogenic (isoprene) VOCs in the ambient air of urban KL during the daytime were determined in this study. ΣBTEX and isoprene averages during the sampling period were 93.25 ± 42.26 µg/m^3 and 6.70 ± 1.57 µg/m^3, respectively. The morning T1 weekday rush hour showed higher BTEX compounds emitted to the air compared to the afternoon T3 and T2. The results clearly can be explained by the volume of vehicles releasing BTEX during incomplete engine fuel combustion. Biogenic isoprene in urban KL was slightly higher compared to forest areas but lower than palm plantation areas. Isoprene in KL was assumed emitted by surrounding palm plantation areas on the outskirts of KL. Both BTEX and isoprene are also ozone precursors that contributed significant OFP. AVOC was found dominant the OFP in the KL urban area, suggesting the impact of anthropogenic activities especially vehicles to the formation of tropospheric ozone.

Acknowledgements

This work was supported by the Universiti Kebangsaan Malaysia (UKM) research grant AP-2015-010 and Ministry of Science, Technology and Innovation (MOSTI) Sciencefund 06-01-02-SF1274.

References

[1] Carlsen L, Bruggemann R and Kenessov B 2018 Use of partial order in environmental pollution studies demonstrated by urban BTEX air pollution in 20 major cities worldwide *Sci. Total Environ.* **610–611** 234–43
[2] Stojić A, Stojić S S, Šoštarić A, Ilić L, Mijić Z and Rajšić S 2015 Characterization of VOC sources in an urban area based on PTR-MS measurements and receptor modelling *Environ. Sci. Pollut. Res.* **22** 13137–52
[3] Kumar A, Singh D, Kumar K, Singh B B and Jain V K 2018 Distribution of VOCs in urban and rural atmospheres of subtropical India: Temporal variation, source attribution, ratios, OFP and risk assessment *Sci. Total Environ.* **613–614** 492–501
[4] Guenther A B, Zimmerman P R, Harley P C and Monson R K 1993 Isoprene and Monoterpene Emission Rate Variability' Model Evaluations and Sensitivity Analyses *J. Geophys. Res.* **98617** 609–12
[5] Fanizza C, Incoronato F, Baiguera S, Schiro R and Brocco D 2014 Volatile organic compound levels at one site in Rome urban air *Atmos. Pollut. Res.* **5** 303–14
[6] Misztal P K, Nemitz E, Langford B, Di Marco C F, Phillips G J, Hewitt C N, MacKenzie A R,

Owen S M, Fowler D, Heal M R and Cape J N 2011 Direct ecosystem fluxes of volatile organic compounds from oil palms in South-East Asia *Atmos. Chem. Phys.* **11** 8995–9017

[7] Li N, He Q, Greenberg J, Guenther A, Li J, Cao J, Wang J, Liao H, Wang Q and Zhang Q 2018 Impacts of biogenic and anthropogenic emissions on summertime ozone formation in the Guanzhong Basin, China *Atmos. Chem. Phys.* **18** 7489–507

[8] Henninger S 2012 Biogenic Isoprene and Its Impact on Human Health in Dependence on Meteorological Conditions -Special Issue "Air Pollution" *J. Environ. Prot. (Irvine,. Calif).* **3** 1206–12

[9] Hallquist M, Wenger J C, Baltensperger U, Rudich Y, Simpson D, Claeys M, Dommen J, Donahue N M, George C, Goldstein A H, Hamilton J F, Herrmann H, Hoffmann T, Iinuma Y, Jang M, Jenkin M E, Jimenez J L, Kiendler-Scharr A, Maenhaut W, McFiggans G, Mentel T F, Monod A, Prévôt A S H, Seinfeld J H, Surratt J D, Szmigielski R and Wildt J 2009 The formation, properties and impact of secondary organic aerosol: current and emerging issues *Atmos. Chem. Phys.* **9** 5155–236

[10] Von Schneidemesser E, Monks P S, Gros V, Gauduin J and Sanchez O 2011 How important is biogenic isoprene in an urban environment? A study in London and Paris *Geophys. Res. Lett.* **38** 1–7

[11] Acton W J F, Schallhart S, Langford B, Valach A, Rantala P, Fares S, Carriero G, Tillmann R, Tomlinson S J, Dragosits U, Gianelle D, Hewitt C N and Nemitz E 2016 Canopy-scale flux measurements and bottom-up emission estimates of volatile organic compounds from a mixed oak and hornbeam forest in northern Italy *Atmos. Chem. Phys.* **16** 7149–70

[12] Hellén H, Tykkä T and Hakola H 2012 Importance of monoterpenes and isoprene in urban air in northern Europe *Atmos. Environ.* **59** 59–66

[13] Molina L T, Molina M J, Slott R S, Kolb C E, Gbor P K, Meng F, Singh R B, Galvez O, Sloan J J, Anderson W P, Tang X, Hu M, Xie S, Shao M, Zhu T, Zhang Y H, Gurjar B R, Artaxo P E, Oyola P, Gramsch E, Hidalgo D and Gertler A W 2004 Air Quality in Selected Megacities *J. Air Waste Manage. Assoc.* **54** 1–73

[14] Rantala P, Järvi L, Taipale R, Laurila T K, Patokoski J, Kajos M K, Kurppa M, Haapanala S, Siivola E, Petäjä T, Ruuskanen T M and Rinne J 2016 Anthropogenic and biogenic influence on VOC fluxes at an urban background site in Helsinki, Finland *Atmos. Chem. Phys.* **16** 7981–8007

[15] Truc V T Q and Kim Oanh N T 2007 Roadside BTEX and other gaseous air pollutants in relation to emission sources *Atmos. Environ.* **41** 7685–97

[16] Lan T T N and Binh N T T 2012 Daily roadside BTEX concentrations in East Asia measured by the Lanwatsu, Radiello and Ultra I SKS passive samplers. *Sci. Total Environ.* **441** 248–57

[17] Norbäck D, Hashim J H, Hashim Z and Ali F 2017 Volatile organic compounds (VOC), formaldehyde and nitrogen dioxide (NO2) in schools in Johor Bahru, Malaysia: Associations with rhinitis, ocular, throat and dermal symptoms, headache and fatigue *Sci. Total Environ.* **592** 153–60

[18] Alghamdi M A, Khoder M, Abdelmaksoud A S, Harrison R M, Hussein T, Lihavainen H, Al-Jeelani H, Goknil M H, Shabbaj I I, Almehmadi F M, Hyvärinen A P and Hämeri K 2014 Seasonal and diurnal variations of BTEX and their potential for ozone formation in the urban background atmosphere of the coastal city Jeddah, Saudi Arabia *Air Qual. Atmos. Heal.* **7** 467–80

[19] Duane M, Poma B, Rembges D, Astorga C and Larsen B R 2002 Isoprene and its degradation products as strong ozone precursors in Insubria, Northern Italy *Atmos. Environ.* **36** 3867–79

[20] Hong Phuc N and Thi Kim Oanh N 2018 Determining factors for levels of volatile organic compounds measured in different microenvironments of a heavy traffic urban area *Sci. Total Environ.* **627** 290–303

[21] Tunsaringkarn T, Prueksasit T, Morknoy D, Semathong S, Rungsiyothin A and Zapaung K 2014 Ambient air's volatile organic compounds and potential ozone formation in urban area,

Bangkok, Thailand *J. Environ. Occup. Sci.* **3** 130–5

[22] Huong Giang N T and Kim Oanh N T 2014 Roadside levels and traffic emission rates of PM2.5 and BTEX in Ho Chi Minh City, Vietnam *Atmos. Environ.* **94** 806–16

[23] Gee I L and Sollars C J 1998 Ambient air levels of volatile organic compounds in Latin American and Asian cities *Chemosphere* **36** 2497–506

[24] MacKenzie A R, Langford B, Pugh T A M, Robinson N, Misztal P K, Heard D E, Lee J D, Lewis A C, Jones C E, Hopkins J R, Phillips G, Monks P S, Karunaharan A, Hornsby K E, Nicolas-Perea V, Coe H, Gabey A M, Gallagher M W, Whalley L K, Edwards P M, Evans M J, Stone D, Ingham T, Commane R, Furneaux K L, McQuaid J B, Nemitz E, Seng Y K, Fowler D, Pyle J A and Hewitt C N 2011 The atmospheric chemistry of trace gases and particulate matter emitted by different land uses in Borneo *Philos. Trans. R. Soc. B Biol. Sci.* **366** 3177–95

[25] Saito T, Yokouchi Y, Kosugi Y, Tani M, Philip E and Okuda T 2008 Methyl chloride and isoprene emissions from tropical rain forest in Southeast Asia *Geophys. Res. Lett.* **35** 2–7

Table 1. Concentration (μg/m^3) of isoprene and BTEX during 5 days sampling T1, T2 and T3 in sampling site.

	30/1/2018 (Tue)			31/1/2018 (Wed)			1/2/2018 (Thu)			2/2/2018 (Fri)			3/2/2018 (Sat)		
Compound	T1	T2	T3	T1	T2	T3	T1	T2	T3	T1	T2	T3	T1	T2	T3
Isoprene	10.52	7.32	7.10	6.79	6.87	5.81	9.02	8.46	6.72	3.28	7.57	5.97	7.00	5.02	3.05
Benzene	25.42	9.23	7.13	12.39	5.43	6.80	17.71	2.58	11.41	20.71	9.83	3.11	3.85	3.00	2.76
Toluene	67.08	42.23	43.72	38.45	26.91	36.94	55.59	16.62	53.50	56.13	37.68	20.90	12.71	10.78	6.32
Ethylbenzene	19.30	13.52	11.95	10.61	7.31	9.56	16.41	5.84	18.13	20.60	13.06	4.32	6.23	5.88	5.22
m,p Xylene	46.14	17.74	17.67	24.09	7.79	12.39	36.41	9.87	29.96	34.68	17.35	13.46	11.14	5.29	3.86
o-Xylene	39.47	17.63	18.40	22.37	12.15	15.03	31.86	9.01	25.09	32.57	18.47	11.47	10.52	6.52	5.48

Table 2. MIR value, daytime average concentration and OFP value during sampling period

		30/1/2018 (Tue)		31/1/2018 (Wed)		1/2/2018 (Thu)		2/2/2018 (Fri)		3/2/2018 (Sat)	
Compound	MIR	μg/m^3	OFP	μg/m^3	OFP	μg/m^3	OFP	μg/m^3	OFP	μg/m^3	OFP
Isoprene	10.68	8.31	88.78	6.49	69.31	8.06	86.13	5.61	59.89	5.03	53.68
Benzene	0.69	13.93	9.61	8.21	5.66	10.57	7.29	11.22	7.74	3.20	2.21
Toluene	3.88	51.01	197.92	34.10	132.30	41.90	162.58	38.24	148.36	9.94	38.57
Ethylbenzene	2.93	14.92	43.73	9.16	26.83	13.46	39.44	12.66	37.10	5.78	16.93
m,p Xylene	7.6	27.18	206.59	14.76	112.17	25.41	193.14	21.83	165.92	6.76	51.38
o-Xylene	7.44	25.17	187.25	16.52	122.87	21.99	163.59	20.83	155.00	7.51	55.86
∑ OFP			733.87		469.14		652.17		574.00		218.61

MIR (Maximum Incremental Reactivity) based on updated value by CARB, 2009
OFP (Ozone Formation Potential) = concentration (μg/m^3) x MIR

Evaluation of Air-Gap Stacked Capacitive Antennas for Lightning Remote Sensing

B. Y. Seah[1], M. R. Ahmad[1], Y. J. Ong[1], N. A. Shairi[1], D. Periannan[1], M. H. M. Sabri[1], M. R. M. Esa[2], Z. Abdul-Malek[2], S. A. Mohammad[2], M. Z. A. A. Aziz[1], N. Yusop[1,4], M. M. Ismail[1], G. Lu[3], V. Cooray[5], A. A. Alkahtani[6] and M. Z. A. Ab. Kadir[6]

[1]Atmospheric and Lightning Research Lab, Centre for Telecommunication Research and Innovation (CeTRI), Fakulti Kejuruteraan Elektronik dan Kejuruteraan Komputer, Universiti Teknikal Malaysia Melaka, Hang Tuah Jaya, 76100 Durian Tunggal, Melaka, Malaysia.
[2]Institute of High Voltage and High Current (IVAT), School of Electrical Engineering, Faculty of Engineering, Universiti Teknologi Malaysia (UTM), 81310 Skudai, Johor Bharu, Malaysia.
[3] Institute of Atmospheric Physics, Chinese Academy of Sciences, China.
[4]Space Science Centre (ANGKASA), Institute of Climate Change, Universiti Kebangsaan Malaysia, 43600 Bangi, Selangor, Malaysia.
[5]Ångström Laboratory, Division for Electricity, Department of Engineering Sciences, Uppsala University, Box 534, S-75121, Sweden.
[6]Universiti Tenaga Nasional (UNITEN), Jalan IKRAM-UNITEN, 43000 Kajang, Selangor.

Email: seahboonyork@yahoo.com; riduan@utem.edu.my

Abstract. This paper evaluates the capacitive antenna performance (sensitivity and frequency shifting) as a lightning sensor in different antenna design structures. Two experiments were carried out, Experiment A – using two different structure antennas, an A3 size capacitive antenna and a quarter A4- sized capacitive antenna, capturing the electric field (E-field) generated by the small spark at a distance of 1 meter away from both antennas. The results were taken in a ratio form of the 1st pulse amplitude captured by quarter A4 size antenna verses 1st pulse amplitude of the waveform captured by A3 antenna for each corresponding spark. Same set up and measurement were taken during the Experiment B except for the quarter A4 size antenna was now being slotted with 6 copper plates were slotted within the gap between the top plate and bottom plate. The result from Experiment A showed an average ratio at 0.1957 with a range of 0.0333 to 0.3085 while the results of Experiment B showed the average value laid at 0.2606 within the range of 0.1581 to 0.4510. The comparison results of Experiment A with Experiment B gave an increment of the sensitivity of the quarter A4 antenna from no stack to fully stacked antenna. However, there was also a significant frequency shifting of the quarter A4 antenna with the change of antenna structure (15% for quarter A4 capacitive antenna without stack; 20% for quarter A4 stacked capacitive antenna).

Content from this work may be used under the terms of the Creative Commons Attribution 3.0 licence. Any further distribution of this work must maintain attribution to the author(s) and the title of the work, journal citation and DOI.
Published under licence by IOP Publishing Ltd

IOP Conf. Series: Earth and Environmental Science **228** (2019) 012002 doi:10.1088/1755-1315/228/1/012002

1. Introduction

Lighting studies had been carried out ages ago by using various types of antennas as the lightning sensor or detector such as helical antenna, dipole antenna, parabolic antenna, ceramic patch antenna and so on to capture the radiation produced by lightning [1, 2, 3, 4, 7, 11]. Field mill, whip antennas and capacitive antennas were commonly used in lightning remote sensing system due to their omnidirectional receiving area and frequency independence properties [6, 7, 8, 9, 10, 13]. These sensors are mainly focused on capturing the vertical components due to the propagation of the E-field produced by a lightning discharge tends to be in vertical manner as based on Maxwell's Ampere Law, the direction of electric field is always following the direction of the current flow (atmospheric sign convention) [13]. However, the flat parallel plates antennas used in previous studies were costly and limited its flexibility to be portable. Alternatively, FR4 was chosen to make the parallel plates antenna due to its cheap production cost and the flexibility of manufacturing in such a various shape of the conducting areas [12]. Throughout the previous studies, it shows that the larger the common area of the capacitive antenna, the higher the sensitivity of the antenna [5]. However, the authors were constructing the antenna in such a size (quarter A4) that is smaller than previous antenna (A3 and A4) in order to reduce the space occupied. This leads to the unidentified of the sensitivity and the frequency shifting (compared to A3 size antenna) of the quarter A4 size antenna. Therefore, the authors would like to study the efficiency of the quarter A4 parallel plates antenna (FR4) with and without stacks compared to A3 parallel plates antenna (FR4).

2. Methodology

Two experiments were conducted where the observations on the derivative electric field change (dE/dt) of antenna without stacks and multi-stacked antenna respectively. An A3-size FR4 antenna was set up as the standard value of the derivative field change to the quarter A4 size antenna throughout these two experiments.

In Experiment A, a quarter A4-sized FR4 antenna with the dimensions of 10cm x 15cm with 3.5cm air gap (as shown in Figure 2) was constructed by having both top and the bottom plates connected to the BNC port via single core wire. The bug killer was constructed with 2 mm gap to produce a small spark, acting as the electromagnetic (EM) wave source throughout the experiment. The distance between antennas and the source of the spark was set to 1 meter to ensure the EM wave generated by the spark will beam at both antennas simultaneously. The setup of the measurement was shown in Figure 4. This FR4 antenna (quarter A4 size) was then slotted with 6 pieces of copper plates with an equivalent gap size of 0.5cm each within its air gap during Experiment B.

Figure 1: Example of copper plates

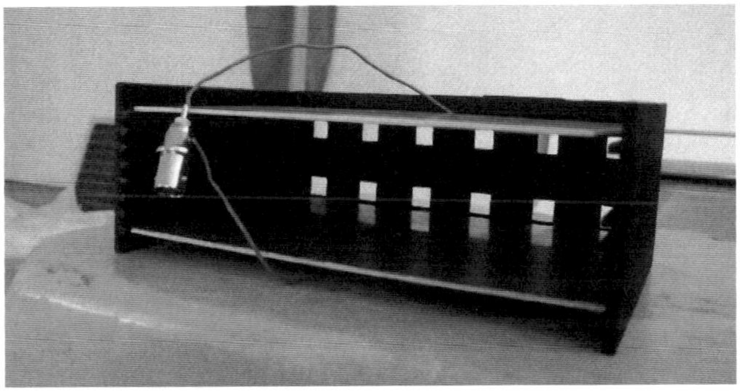

Figure 2: Example of quarter A4 size antenna (without stack)

Figure 3: Example of quarter A4 size antenna (fully stacked)

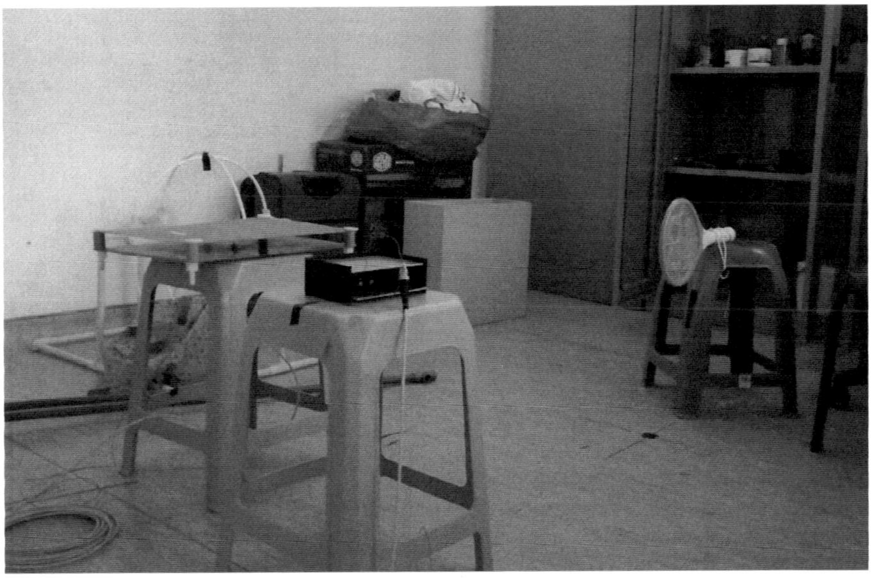

Figure 4: Measurement setup of Experiment A and B

IOP Conf. Series: Earth and Environmental Science **228** (2019) 012002 doi:10.1088/1755-1315/228/1/012002

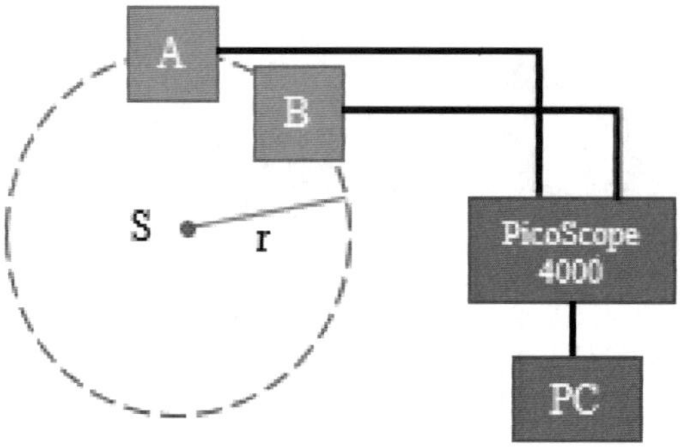

Figure 5: Schematic diagram of the measurement setup in Experiment A and Experiment B. (A: quarter A4 size antenna; B: A3 size capacitive antenna; S: small spark source)

3. Results, analysis and discussion

The results obtained were analysed from two aspects, the performance ratio of the 1st pulse amplitude captured by quarter A4 size antenna verses 1st pulse amplitude of the waveform captured by A3 antenna and the presence of frequency shifting for each corresponding spark.

Note that the statistical distribution of the ratio which studied about the sensitivity of the antenna with different structure were done by taking the first ten samples with slightly and very slightly frequency shifting while the frequency shifting of the quarter A4 antenna was taken from the first twenty samples from each experiment and classified into four classes as shown in Table 1. *The first ten samples used in the analysis of performance ratio are included in the first twenty samples.

Table 1: Classification of the frequency shifting property of quarter A4 parallel plates antenna

Class	Description
Yes	Different onset polarity; the polarity of each gradient within 1st pulse of quarter A4 sized and A3 sized antennas were not same.
Multipeak	Same onset polarity but drastic changes in gradients' polarities before the end of the 1st pulse.
Slightly	Same onset polarity and each gradient within 1st pulses but different polarity of the gradient after 1st pulse.
Very Slightly	Same polarity for onset, and each gradient of increment or decrement throughout the whole discharge waveform.

The detail of the waveform for each types of classification were shown in Figure 6 until Figure 9 which were taken directly from the PicoScope software (Note that the blue colour is the waveform captured by A3 capacitive antenna while the red colour is the waveform captured by quarter A4 capacitive antenna). As shown in these figures, "Yes" class shows the waveforms with different onset polarity. These might be due to there were some frequency components that could only detect by an antenna with larger dimension instead. These abilities to capture the missing frequency components is one of the factors affecting the efficiency and sensitivity of the antenna itself. As for the Figure 8, it shows a similar waveform for both antennas but there is a slightly different in the polarity of the gradient at the end part due to unknown reason. These cases were classified as "Slightly" frequency shifting waveforms. For the "very slightly" class signals, their gradient at each particular point are the same slope polarity although there are different in the E-field changes amplitudes.

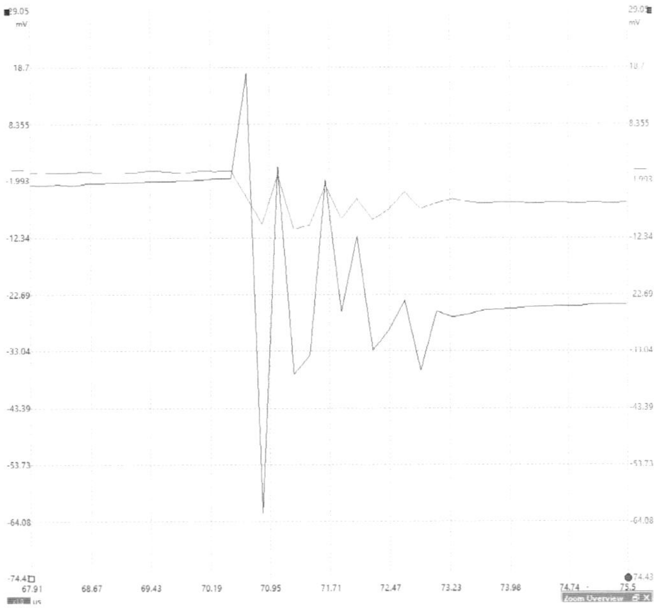

Figure 6: Example of "Yes" classical waveform

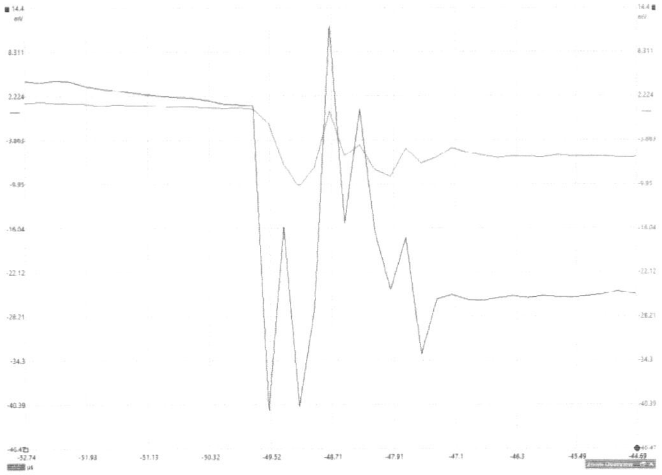

Figure 7: Example of "Multipeak" classical waveform

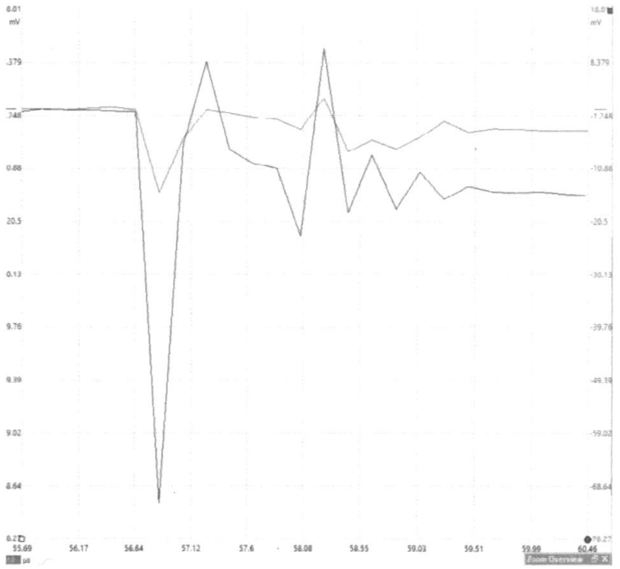

Figure 8: Example of "Slightly" classical waveform

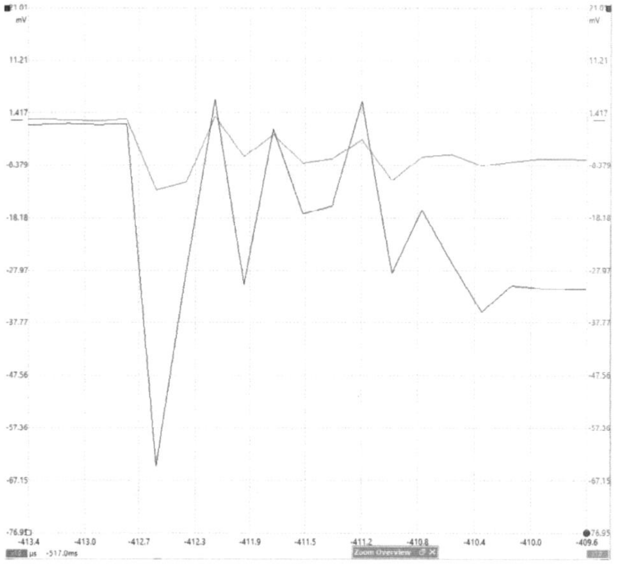

Figure 9: Example of "V. slightly" classical waveform

Figure 10 shows the amplitude ratio of quarter A4 without stack to A3 antenna which have the similar waveform of their first pulse (same polarity). Comparing Figure 10 and Figure 11 with Figure 12 and Figure 13 respectively, it shows that stacking the air gap of the parallel plates antenna with copper plates will have an increment in its sensitivity about 33.16 percent in average, but this will also lead to a frequency shifting to the antenna. Based on the description in Table 1, we know that the signals with very slightly frequency shifting could be consider as no shifting as their signals waveforms are nearly the same. There are ten samples out of twenty samples were found to be very similar (very slightly waveforms) captured by the quarter A4 antenna without stack. However, this number decreases to seven only when the antenna was stacked with 6 pieces of copper plates. The overall number of signals with significant frequency shifting by one sample per twenty samples.

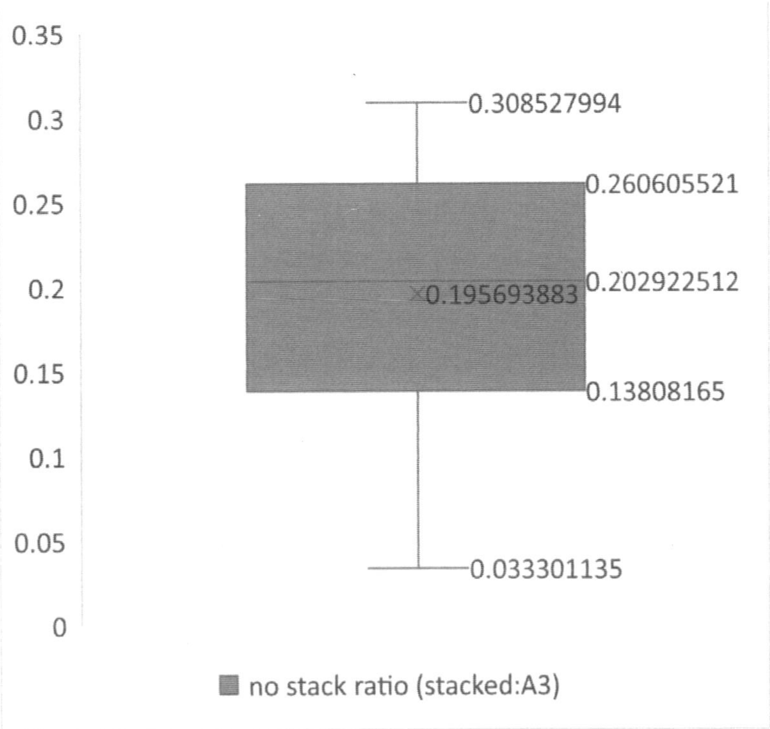

Figure 10: Statistical distribution of the ratio of the 1st pulse amplitude captured by quarter A4 size antenna without stack verses 1st pulse amplitude of the waveform captured by A3 antenna for each corresponding spark

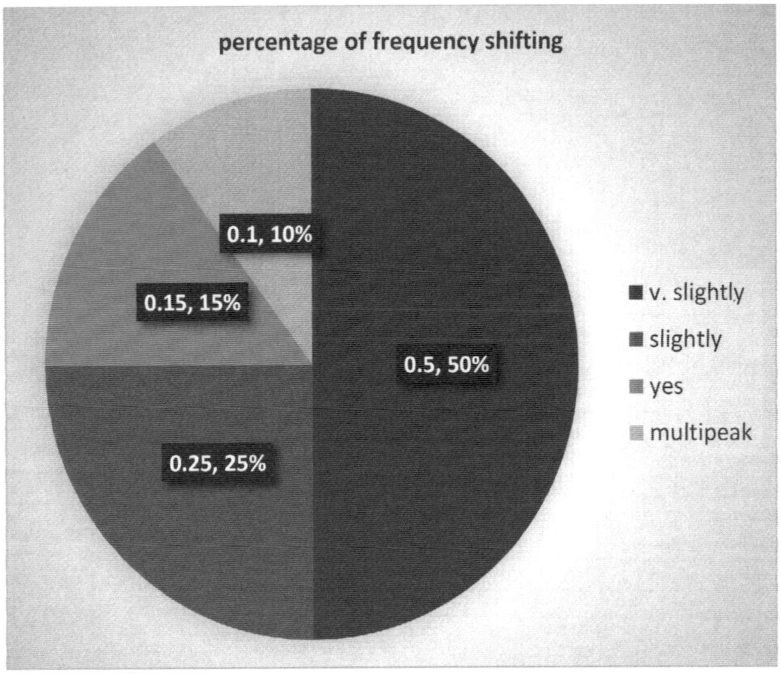

Figure 11: Percentage of frequency shifting of quarter A4 antenna without stack

Unlike the antenna without stacked, the statistical distribution of the stacked antenna shows a skewed distribution among the twenty samples taken. One of the samples proves that stacking the antenna will enable the antenna to reach nearly half of the sensitivity of A3 size capacitive antenna although it had been improved about one-third of the performance from quarter A4 antenna without stacks.

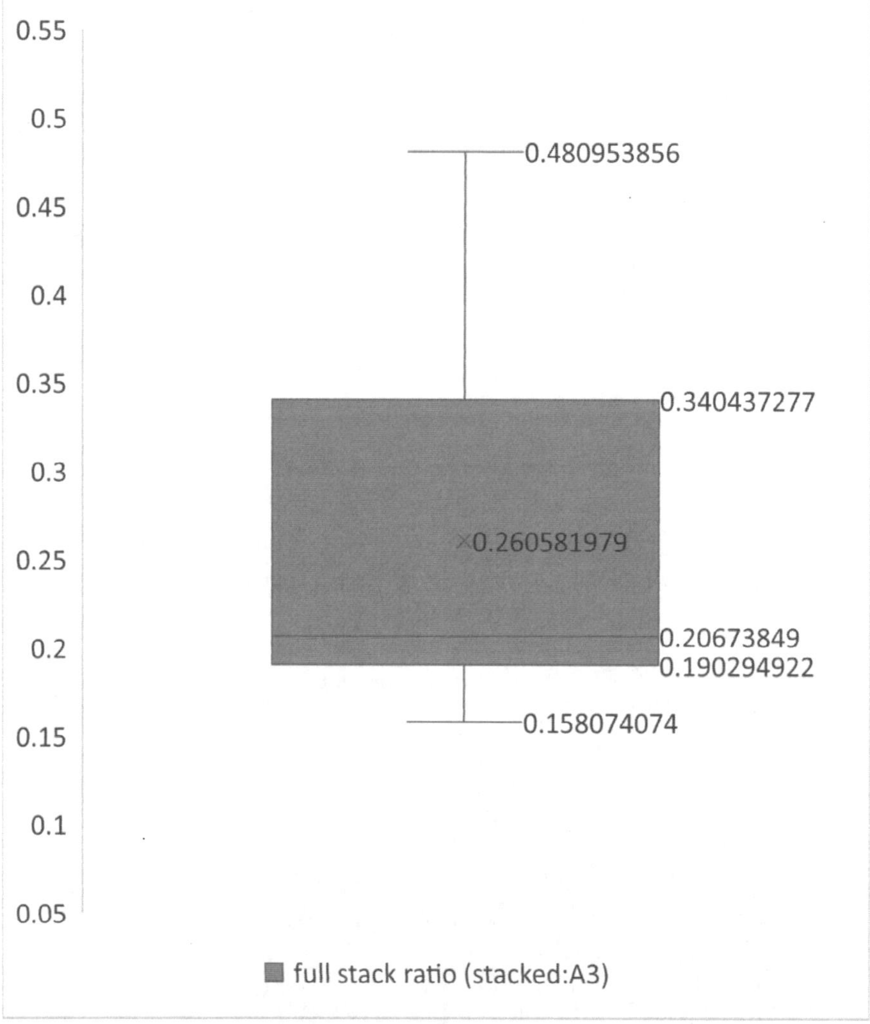

Figure 12: Statistical distribution of the ratio of the 1st pulse amplitude captured by fully stacked quarter A4 size antenna verses 1st pulse amplitude of the waveform captured by A3 antenna for each corresponding spark

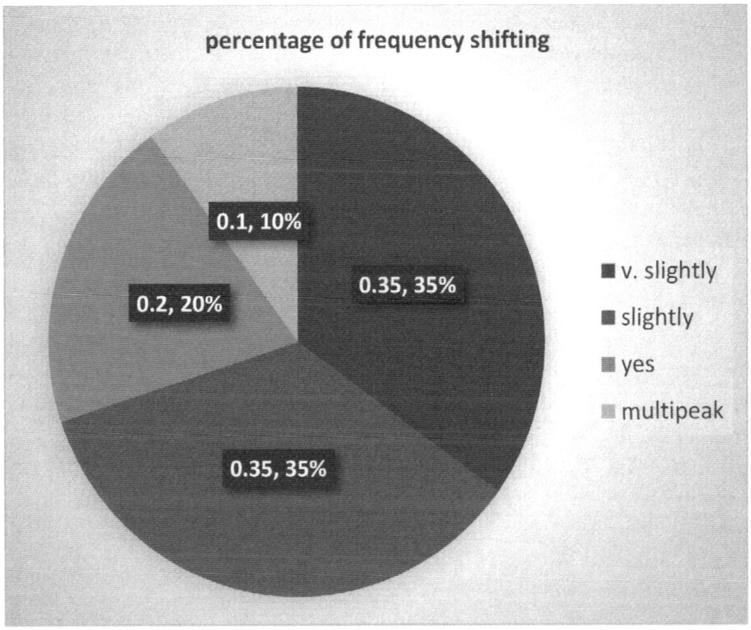

Figure 13: Percentage of frequency shifting of the fully stacked quarter A4 antenna

4. Conclusion

The authors have analysed the ratio of the 1st pulse amplitude of the EM waves from small spark by considering the first ten samples with Slightly and Very slightly frequency shifting properties that captured by the quarter A4 antenna for Experiment A and Experiment B. The waveforms (captured by quarter A4 antenna) were classified into four classes, Yes, Multipeak, Slightly, and Very slightly for their frequency shifting properties. As a conclusion, the smaller the size of the antenna, the lower the sensitivity of the antenna itself. The presence of the stack components (copper plates) will not only increase the sensitivity (33.16%) but also the percentage of frequency shifting of the antenna (33.33%). Studies in detail about the performance of the smaller size antenna could be done by normalizing and analyzing the E-field change captured by both antennas (A3 and quarter A4) in future.

Acknowledgments

The authors would like to acknowledge the support provided by Fakulti Kejuruteraan Elektronik dan Kejuruteraan Komputer (FKEKK), Universiti Teknikal Malaysia Melaka (UTeM). This project is funded by Short Term Grant (Projek Jangka Pendek PJP) in FKEKK, UTeM (PJP/2018/FKEKK/(3B)/S01615). This project is partially funded by Universiti Teknologi Malaysia (UTM) under Vote No. 04G19 and Ministry of Education (MOE) under 4F966 research grants (FRGS).

References

[1] Ligda, M.G., 1956. *The radar observation of lightning.* Journal of Atmospheric and Terrestrial Physics, 9(5-6), pp.329-346.

[2] Brook, M. and Kitagawa, N., 1964. *Radiation from lightning discharges in the frequency range 400 to 1000 Mc/s. Journal of Geophysical Research*, *69*(12), pp.2431-2434.

[3] Kosarev, E.L., Zatsepin, V.G. and Mitrofanov, A.V., 1970. *Ultrahigh frequency radiation from lightnings.* Journal of Geophysical Research, 75(36), pp.7524-7530.

[4] Rust, W.D., Krehbiel, P.R. and Shlanta, A., 1979. *Measurements of radiation from lightning at*

2200 MHz. Geophysical Research Letters, 6(2), pp.85-88.

[5] Galvan, A. and Fernando, M., 2000. *Operative characteristics of a parallel-plate antenna to measure vertical electric fields from lightning fields from lightning flashes. Report UURIE*, pp.285-00.

[6] Sonnadara, D. U. J., Edirisinghe, C. M., & Fernando, I. M. K., 2001. *Construction of a High-Speed Buffer Amplifier to Measure Lightning Generated Vertical Electric Fields*.

[7] S. Yoshida, July 2008. *Radiations in Association with Lightning Discharges*, Department of Information and Communications Technology, Division of Electrical, Electronic and Information Engineering Graduate School of Engineering, Osaka University, Japan.

[8] Esa, M.R.M., Ahmad, M.R., and Cooray, V., 2014a. *Wavelet analysis of the first electric field pulse of lightning flashes in Sweden*. Atmospheric Research, 138, pp.253–267.

[9] Esa, M.R.M., Ahmad, M.R., Rahman, M., and Cooray, V., 2014. *Distinctive features of radiation pulses in the very first moment of lightning events*. Journal of Atmospheric and Solar-Terrestrial Physics, 109, pp.22–28.

[10] Ahmad, M.R., Esa, M.R.M., Cooray, V. and Dutkiewicz, E., 2014. *Interference from cloud-to-ground and cloud flashes in wireless communication system*. Electric Power Systems Research, 113, pp.237-246.

[11] Petersen, D. and Beasley, W., 2014. *Microwave radio emissions of negative cloud-to-ground lightning flashes*. *Atmospheric research*, *135*, pp.314-321.

[12] Dinesh a/l Periannnan, 2016. *Synthesis of Capacitive and Inductive Antennae Designed for Lightning Sensor Application*. Undergraduate research thesis, FKEKK, UTeM, Malacca, Malaysia.

[13] Cooray, V., 2015. *An introduction to lightning*. Springer.

Bromocarbons over the Coastal Area in Peninsular Malaysia

M. S. Mohd Nadzir[1,2], S. N. Mohd Yusof[1] and M. T. Latif[1]

[1]School of Environmental and Natural Resources Sciences, Faculty of Science and Technology, Universiti Kebangsaan Malaysia, 43600 UKM Bangi, Selangor, Malaysia.
[2]Centre for Tropical Climate Change System (IKLIM), Institute for Climate Change, Universiti Kebangsaan Malaysia, 43600 UKM Bangi, Selangor, Malaysia.

shahrulnadzir@ukm.edu.my

Abstract. Concentration of bromoform ($CHBr_3$), dibromochloromethane ($CHBr_2Cl$) and dichlorobromomethane ($CHBrCl_2$) in seawater was analysed to determine the concentrations at Cape Rachado, Negeri Sembilan (mangrove area) and Mėlawi Beach, Kelantan (coastal area) and factors that can influenced the concentrations. Seawater samples have been identified and quantify by using purge-and-trap concentrator (P&T) and gas chromatography equipped with electron captured detector (GC/ECD). Mean value of CHBr3 at Cape Rachado is 0.010 mmol/L and Melawi Beach is 0.001 mmol/L. The major brominated halocarbons at Cape Rachado and Melawi Beach is $CHBr_3$ with range value 0.009 – 0.012 mmol/L and 0.0004 – 0.0021 mmol/L, respectively. Concentrations of $CHBr_3$, $CHBr_2Cl$ and $CHBrCl_2$ at Cape Rachado (mangrove area) are higher than Melawi Beach (coastal area) because the located at seaweed bed and adjacent to mangrove swamp.

1. Introduction

Emissions of biogenic halocarbons showed large variability and mainly based on natural oceanic origin, especially in tropical coastal such as salt marshes and mangrove that locates large diversity of phytoplankton and macroalgae [1][2]. Study on bromocarbons from tropical latitudes ocean critically important because of deep convection allows for a rapid ascent of the bromine compounds from the marine boundary layer to tropical tropospheric layer (TTL). Also, transport pathways from marine boundary layer to the TTL are important to identify for assessing stratospheric ozone depletion [3]. Emphasis on bromocarbons study is because as they are readily oxidised and formed highly ozone-depleting inorganic bromine compounds [4].

The goal of this study is to determine the concentration of bromocarbons ($CHBr_3$, $CHBr_2Cl$ and $CHBrCl_2$) at different areas (coastal and mangrove area) by using purge-and-trap concentrator (P&T) and gas chromatography with electron captured detector (GC/ECD) and identify the factors that contribute to the presence.

Content from this work may be used under the terms of the Creative Commons Attribution 3.0 licence. Any further distribution of this work must maintain attribution to the author(s) and the title of the work, journal citation and DOI.
Published under licence by IOP Publishing Ltd

IOP Conf. Series: Earth and Environmental Science **228** (2019) 012003 doi:10.1088/1755-1315/228/1/012003

2. Methodology

2.1 Study site

Water samples were collected from Cape Rachado, Negeri Sembilan (area A) and Eastern Coast, Melawi Beach, Bachok, Kelantan (area B) during North-East Monsoon (November to March). Both areas has two monsoon seasons, north-east and south-west with mean annual rainfall about 2,500 mm, range of temperature is from 25°C to 32°C and wind flow 10 to 20 knots.

Cape Rachado (2°24'54.63" N, 101°51'16.21" E) is located in the south-western region of the state of Negeri Sembilan, Malaysia, facing Straits of Malacca and approximately about 18 km south from Port Dickson town. This nature attraction becomes favourite spot for bird watching during migratory season of raptors, also have hidden beaches, mangrove swamp and famous picnic area. Area A has a fringing coral reef flats extend out about 50 m, gently slope off about 8 m towards the reef edge [5]. The reef flats were completely exposed when the tidal level is about less than 0.1 m from chart datum [6].

Melawi Beach (6°0'33.77" N, 102°25'37,64" E) is located 37 km south from Kota Bharu town, capital state of Kelantan or approximately 45 min driving by car and facing South China Sea (SCS). Famous location among tourists and local peoples because it located away from the bustling town and many accommodations was built near Melawi Beach. Other than that, Melawi Beach is adjacent to a small port for fishing, Tok Bali which has served as jumped-off point for tourists to go to Perhentian Islands. Sampling locations are shown in Figure 1.

2.2 Sampling technique

Collections of water sample were taken from each area according to standard methodologies, during daylight and low tidal level with maximum level of 0.8 m. Niskin sampler 4 L (OSIL, UK) was used to collect the samples and placed in Schott bottle wrapped with aluminum foil without headspace. Collected samples were stored in lightproof insulated box with ice pack and water for rapid cooling before being transported back and stored in chest freezer in laboratory at 4°C for less than one week to prior to analysis. Location was located by GPS and in-situ parameters (water temperature, pH and dissolved oxygen) readings were taken by portable YSI Multi-probe field meter model 556. Replicate of samples not less than three. The study conducted is short term measurement. Sampling details are shown in Table 1.

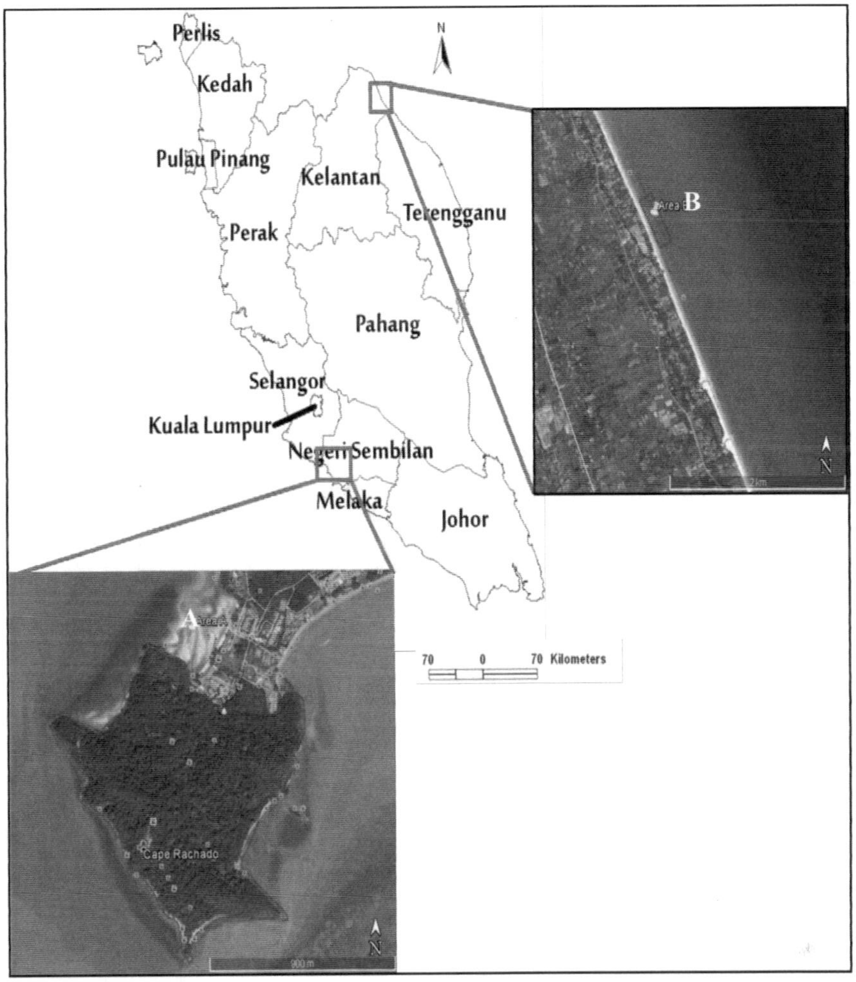

Figure 1. Map of study areas: (a) Cape Rachado and (b) Melawi Beach.

Table 1. Descriptions of at Cape Rachado and Melawi Beach.

Study area	Coordinate	Description
Cape Rachado (area A)	2° 24' 54.63" N 101° 51' 16.21" E	Mangrove swamp, forest, coral reef, seaweed bed, recreation and vacations areas.
Melawi Beach (area B)	6° 0' 33.77" N 102° 25' 37.64" E	Fishing, recreation and vacation areas.

2.3 Instrumentation and analysis

For calibration, a stock solution of a Trihalomethane mixture (Sigma, US) containing each compound at 200 mg/L in methanol were diluted to 5 different concentrations in pure methanol (Merck, Germany) placed in amber bottle and stored at 4°C for less than one week. The internal standard solution consists four compounds, $CHBr_3$, $CHCl_3$, $CHBr_2Cl$ and $CHBrCl_2$ but this study focused to bromocarbons only. Calibration were performed by spiked a few mg/L of mixed standard solution into a system, from low to high concentrations and followed by sample analysis. Each concentration of standard solution and sample was analyzed not less than three times (triplicates). The peak areas for each quantify ions were recorded against known concentration bromocarbons and fitted into linear equation derived to determine the concentrations of compounds.

Bromocarbons analysis was performed with Tekmar Dohrmann 3100 purge-and-trap concentrator (Tekmar, Ohio) coupled with Varian CP-3800 Gas Chromatograph - Electron Captured Detector (Varian,USA). The instrument was controlled and interpreted chromatograph data by GC Workstation Version 6.41 software. A 5 mL filtered water sample was manually injected into purge-and-trap system, purged with pure nitrogen (N_2) gas to extract the volatile compounds at a flow-rate 40 mL/min for 11 min ± 0.1 min at ambient temperature. The extracted & trapped gases moved to desorption stage at temperature of 180°C for 4 min before transfer to GC/ECD for identification and quantification phase.

Identification and quantification of compounds were performed with fitted narrow-bore DB5 column (J&W, USA), 30 m length, 0.32 mm internal diameter and 1.0 µm film thickness. Detector used nitrogen (N2) as make-up gas at 60 mL/min and high purity helium (He) as carrier gas at 2.5 mL/min. GC/ECD was set to temperature program: 40 °C for 5 min, 130 °C at 8 °C/min, 180°C at 25°C/min and held for 5 min for last temperature. The temperature of column, injector and detector was controlled at 40 °C, 200 °C and 250 °C respectively. Total time for analysis samples with P&T and GC/ECD is 41 min.

3 Result and Discussion

The method optimised for detection of bromocarbons ($CHBr_3$, $CHBr_2Cl$ and $CHBrCl_2$) in seawater samples from mangrove and coastal areas. The limit of detection (LOD) for each bromocarbon compound by using GC/ECD are 0.3 fmol/L for $CHBr_3$, 0.2 fmol/L for $CHBr_2Cl$ and 0.4 fmol/L for $CHBrCl_2$. Retention time for $CHBr_3$ is 9.254 min, $CHBr_2Cl$ is 6.637 min and 4.143 min for $CHBrCl_2$. Recoveries obtained for $CHBr_3$ is 84.7 – 89.4%, $CHBr_2Cl$ is 81.5 – 86.6% while $CHBrCl_2$ is 79.4 – 80.7%. The mean value of seawater quality for in-situ parameters are 28.1 – 33.5°C for seawater temperature, 8.71 – 10.50 mg/L for dissolved oxygen and 8.21 – 8.71 for pH.

Result of bromocarbons in seawater samples obtained from sampling at two different areas, mangrove and coastal are showed in Figures (a) and (b), respectively. Range concentrations for each compound from mangrove area are 0.009 – 0.012 mmol/L for $CHBr_3$, 0.005 – 0.006 mmol/L for $CHBr_2Cl$ and 0.006 – 0.008 mmol/L for $CHBrCl_2$. Mean value of $CHBr_3$ is 0.010 mmol/L, highest than mean value of $CHBr_2Cl$ (0.006 mmol/L) and $CHBrCl_2$ (0.007 mmol/L). A presence of bromocarbon compounds in seawater is because the study area is located at seaweed bed and adjacent to small patch of mangrove. Seaweeds or macroalgaes functional as collector that collects halides from seawater uptakes and produce halocarbons by oxidation process [7]. From observation, geographical condition at study area has abundance of lime stones and rocks at sea ground are suitable for macroalgaes to develop and growth [8]. Cape Rachado dominated by the brown seaweed (Sargassum binderi, Turbinaria comoides and Padina australis), followed by green and red seaweed. Brown algae known as major contributor of brominated halocarbons compared with green and red algae [9].

Concentration of $CHBr_3$ in seawater from coastal area still dominated compared with another two compounds, $CHBr_2Cl$ and $CHBrCl_2$. Concentrations range value for bromocarbons from coastal area are 0.0004 – 0.0021 mmol/L for $CHBr_3$, 0.0007 – 0.0012 mmol/L for $CHBr_2Cl$ while for $CHBrCl_2$ is 0.0003 - 0.0004 mmol/L. Mean value of $CHBrCl_2$ (0.0003 mmol/L) is the lowest compared to $CHBr_3$ and $CHBr_2Cl$. Concentration of $CHBr_3$ increased when phytoplankton increased [10]. Based on observation along coastline at Melawi Beach, there is no presence of seaweed bed. Thus, concentrations of bromocarbons at Melawi Beach may not be influenced by seaweed emissions. Detected bromocarbons in seawater could be contributing by microalgae such as phytoplankton [11]. Another factor that could be the influence the bromocarbons concentration is wind direction. Movement of atmospheric bromocarbons can be explained with software NOAA Hybrid Single-Particle Lagrangian Integrated Trajectory (HYSPLIT) model by using 3 days backward trajectory to indicate the transportation of air masses to study area. Figure 3 showed the transportation of air masses during sampling time. The air masses transports from Northeast Peninsular Malaysia bring together the emissions from South China Sea (SCS) before arrived to Melawi Beach. Emissions from SCS and phytoplankton can be the factors in influenced bromocarbons concentrations.

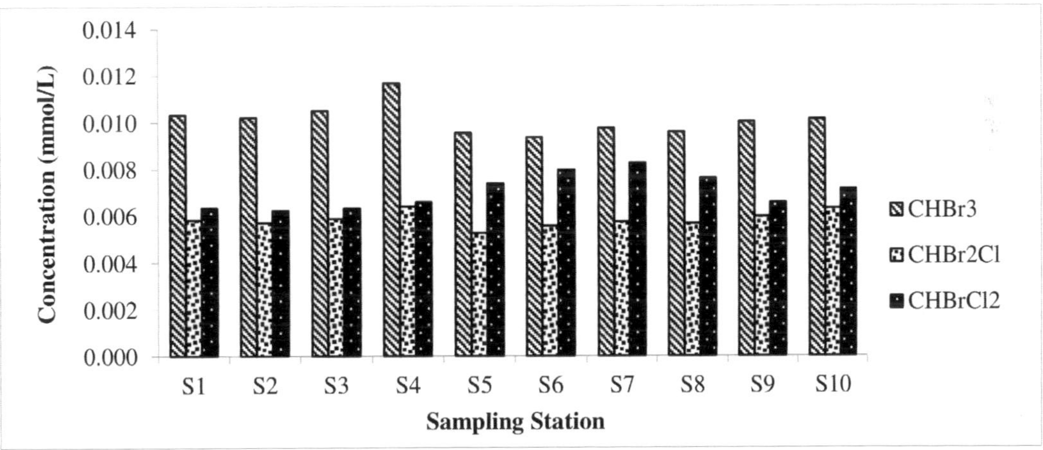

(a)

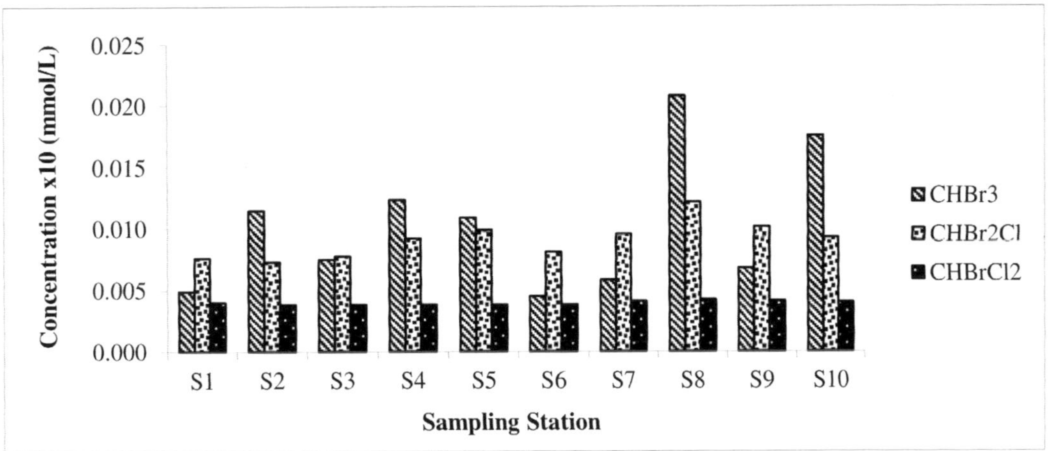

(b)

FIGURE 2. (a) Concentrations of bromocarbons in seawater from mangrove area. (b) Concentrations of bromocarbons in seawater from coastal area.

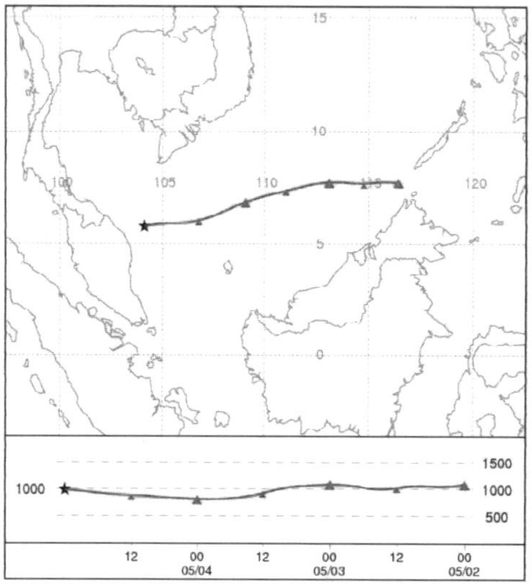

Figure 3. Transportation of air mass by NOAA HYSPLIT 3 days backward trajectories to Melawi Beach.

4. Conclusion

It was concluded that $CHBr_3$ was the dominant bromocarbons at mangrove and coastal area compared to $CHBr_2Cl$ and $CHBrCl_2$. The analysis showed, seaweed and mangrove trees are the sources of $CHBr_3$, $CHBr_2Cl$ and $CHBrCl_2$ in seawater at Cape Rachado, while phytoplankton and air masses from North East Peninsular through South China Sea (SCS) can be sources of bromocarbons at Melawi Beach. Mean value of $CHBr_3$ at Cape Rachado (0.010 mmol/L) is higher than Melawi Beach (0.001 mmol/L). Also, range value for concentrations of bromocarbons at Cape Rachado higher than Melawi Beach with 0.009 – 0.012 mmol/L for $CHBr_3$, 0.005 – 0.006 mmol/L for $CHBr_2Cl$ and 0.006 – 0.008 mmol/L for $CHBrCl_2$.

ACKNOWLEDGMENTS

This study was funded by Universiti Kebangsaan Malaysia (UKM) under research grant number GUP-2014-014, also thank you to School of Environmental Science and Natural Resources and Chemistry Department UKM for instruments facilities.

References

[1] Quack B and Wallace D W D R Wallace, 2004, Global Biogeochemical Cycles. **17**, 23-1:23-27.
[2] Dessens O, Zeng G, Warwick N and Pyle J, 2009. Atmos. Sci. Let. 10, 201-206.
[3] Pyle J A, Ashfold M J, Harris N R P, Robinson A D, Warwick N J, Carver, G D, Gostlow B, O' Brien L M, Manning A J, Phang S M, Yong S E, Leong K P, Ung E H and Ong S, 2011, Atmos. Chem. Phys. **11**, 529-542.
[4] Hosking J S, Russo M R, Braesicke P and Pyle J A, 2012, Atmos. Chem. Phys. 12, 9791-9797.
[5] Goh A H and Sasekumar A, 1980, Malay. Nat. J. 34, 25-37.
[6] Phang S M, 1995, Malaysian Journal of Science. 16A, 23-32.

IOP Conf. Series: Earth and Environmental Science **228** (2019) 012003 doi:10.1088/1755-1315/228/1/012003

[7] Manley S L, Goodwin K D and North W J, 1992, Limnol. Oceanogr. 37, 1652-1659.
[8] Searless R B, 1984, Helgol. Meer. 38, 259-271.
[9] F. Laturnus, T. Svensson and C. Wiencke, Polarres. 29, 379-384 (2010).
[10] J. Karlsson, A. Jonsson, M. Meili and M. Jansson, Limnol. Oceanagr. 48, 269 - 276 (2003).
[11] B. Quack, E. Atlas, G. Petrick, V. Stroud, S. Schauffler and D. W. R. Wallace, Geophys. Res. Lett. 31, L23S05 (2004).

The Performance of Medium Access Control Protocol with Capture Effect for Lightning Remote Sensing

N. Q. Jalaudin[1], M. R. Ahmad[1], M. Z. A. A. Aziz[1], M. R. M. Esa[2]

[1]Broadband and Networking Research Group (BBNET), Centre for Telecommunication Research and Innovation, Fakulti Kejuruteraan Elektronik dan Kejuruteraan Komputer (FKEKK), Universiti Teknikal Malaysia Melaka (UTeM), Hang Tuah Jaya, 76100 Durian Tunggal, Melaka, Malaysia.
[2]Institute of High Voltage and High Current (IVAT), Universiti Teknologi Malaysia (UTM), 81310 Skudai, Johor Bharu, Malaysia

Email: riduan@utem.edu.my, monariza@utm.my

Abstract. In this paper we investigate the capture phenomenon for IEEE 802.11b standard. The proposed model allows us to derive the indoor and outdoor of approximate range in propagation loss and shadowing for various applications in remote sensing and monitoring. We estimate the network performance by considering the path loss and shadowing for two distance values which are 35 m and 140 m. We can conclude that that shorter distance has higher performance in path loss and shadowing than the longer distance. These results can be used as a useful guide to scientist and engineers before the communication network deployed to transfer captured data to the gateway or control centre.

1. Introduction

Medium access control (MAC) is developed for wireless network. The medium accesses are shared, based on the rules of the devices. It can communicate with each other. The multiple devices can access at the same time because the nature of wireless medium [1].

In the early 1970s, wireless MAC protocols have been studied widely. The first method is ALOHA, when a node is ready for a transmission, it transmits the data immediately. Since all nodes share the same radio channel, when two or more nodes begin the transmissions at the same time, collision occurs. All the data transmission cannot be detected and read correctly [2].

Slotted ALOHA is an improvement of ALOHA, the time is divided into slot of sizes. Each node start to transmit frames only at the beginning slot. If two or more frames collide in a slot, then all nodes detect the collision event before the slot ends. The colliding frames are buffered and retransmit after random retransmission delay. It attempts retransmission with probability number between 0 and 1 [3].

The method has been enhanced with the addition of the detection process before transmitting. In order to decrease the collision probability, Carrier Sense Multiple Access (CSMA) was developed. In

Content from this work may be used under the terms of the Creative Commons Attribution 3.0 licence. Any further distribution of this work must maintain attribution to the author(s) and the title of the work, journal citation and DOI.
Published under licence by IOP Publishing Ltd

this method, nodes listen to the medium before transmitting. If it senses idle node, it transmits immediately whereas if it senses busy node, it delays access and retries later [4].

In this paper, we focused on a Carrier Sense Multiple Access with Collision Avoidance (CSMA/CA) protocol. The main purpose of CSMA/CA is to avoid the collision. This method uses handshaking process. In order to reduce the duration of collision, it uses small Request-to-send/Clear-to-send (RTS/CTS) packets exchanged at the basic control rate to reserve the medium before large packets are transmitted [5]. The transmitter sends a RTS to intended receiver, when it has DATA packet to send. All nodes in the transmitter region keep silent, while the receiver sends back a CTS packet to the intended transmitter. All nodes in the receiver region keep silent too. The transmitter sends DATA to the intended receiver, while the receiver replies back with an acknowledgement packet to the transmitter after successfully receives DATA [5].

In this paper we study the performance of IEEE 802.11b Local Area Network/Metropolitan Area Network (LAN/MAN) standard [6]. IEEE 802.11 is the WLAN MAC and Physical Layer (PHY) specifications [7]. In wireless services it has several of services which are support data, voice and video. The advantages of wireless are better mobility, cost effective, VOIP facility and scalability.

This paper is organized as follow. In section II, the model of protocol of CSMA/CA protocol is described. The throughput and average delay of CSMA/CA are analyzed in section III. Lastly, concluding remark is provided in Section IV.

2. Critical Review

There is growing interest in the research community to design and develop model of protocol to examine the performance, potential and operation in communication system. In [11], the authors proposed the influence a capture effect with L-fold antenna diversity at the access point over IEEE 802.11b DCF at data rate 11Mbit/s. In this paper it designed in IEEE 802.11b Basic Service Set (BSS) for indoor communication. In order to theoretically predict the saturation throughput of the single IEEE 802.11b BSS at 11Mbit/s under influence of the capture effect in Rayleigh, the author applied obtained analytical results for the maximum selection and the ideal selection diversity strategy. It compared with the capture model non diversity, capture model with 2-fold maximum selection and 2-fold ideal selection diversity.

In [12], the authors proposed to achieve quality-of-service (QoS) differentiation in IEEE 802.11 wireless LAN in dual transmission power. In this paper it designed the QoS differentiation in IEEE 802.11 wireless LAN by allocating a higher transmission power to hosts requiring a higher service priority, where compare with high-power transmissions have a higher chance and low-power ones to capture the channel in the event of mutual collision. Lastly, the author confined their analysis to single-cell system because the effect of using several transmission power levels in a multiple cell context.

3. Protocol model

Communication channel is the path that data takes from transmitter to the destination. Communication channel also known as transmission media or communication media. Communication channel is divided into types which are guided (wired) and unguided (wireless).

In wired communication information is transferred between two or more nodes using wires while in wireless communication information is transferred between two or more nodes that are not connected using electrical conductor. An information is transmitted via open space known as radio waves [8].

Radio waves are generated by radio transmitter and received at the radio receiver. During the transmission from a transmitting antenna to a receiver, path loss occurs due to the propagation medium. The basic propagation loss is shown in Figure 1.

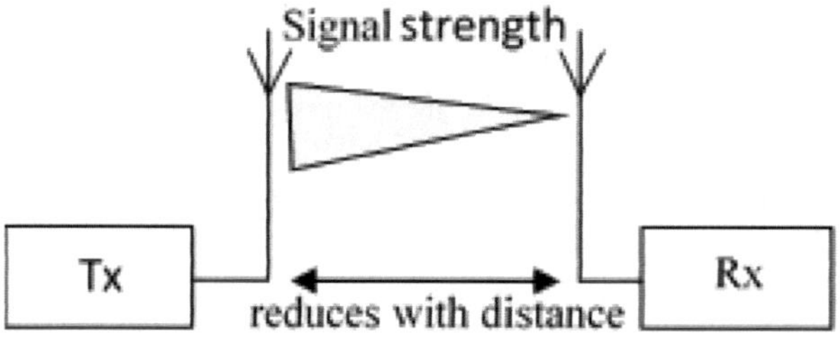

Figure 1. Basic Propagation loss

Propagation loss can attenuate the transmitted signal. In general, to measure the signal strength transmitted packet decrease with the distance between the access point and node is increase.
The propagation loss model can be categorized into two types which are indoor environment and outdoor environment. In indoor environment, the signal strength is much smaller range of access point and other node separation and the variability of environment are much greater and vice versa in outdoor environment.

The effect that the received signal power fluctuates due to the obstructing the propagation path between transmitter and receiver is known as shadowing [9].
In the propagation path, the signal strength of a radio signal is attenuated by obstructions. By using the same frequency, the node with stronger received signal can successfully transmit its packet to the access point even though two or more nodes simultaneously transmit their packet. This scenario is called capture effect [10].

In general, if two signals are received on the same channel, only the stronger of the two appears in the output. To allow the stronger signal to capture the channel, the threshold will be set at the receiver limiter called capture ratio. In this work, we present a communication channel and approximately derive capture effect, which is defined as the channel successful transmit the packet, propagation loss and shadowing. The model of protocol is shown in Figure 2.

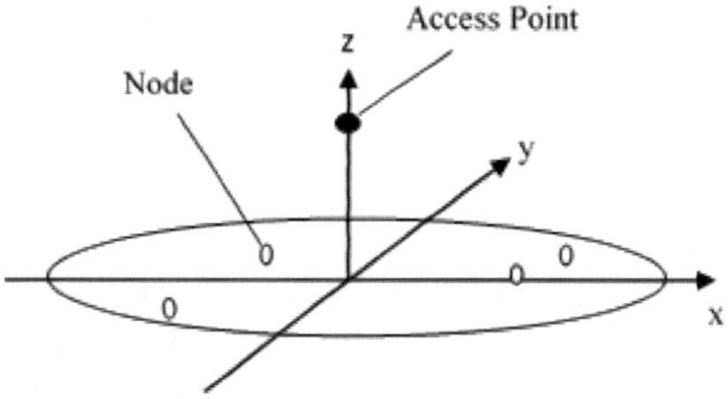

Figure 2. Model of protocol

IOP Conf. Series: Earth and Environmental Science **228** (2019) 012004 doi:10.1088/1755-1315/228/1/012004

4. Performance analysis

The modification of the traditional CSMA protocol, the CSMA/CA has been proposed by adopting the concept of back off time to perform collision avoidance. In this scheme, is to solve the hidden problem by informing the communication channel conditions from the access point to the users.

Based on the algorithm of CSMA/CA, each node must wait a period of time is known as backoff time. When the access point receives packets from the user it will sends a busy signal to all users. On the other hands, if the access point not receiving any packet it will send idle signal. In IEEE 802.11 standard, random backoff time is shown as BackoffTime = Random() x SlotTime. Where random is represents the pseudorandom integer [0, CW]. The CW is an integer within the range of value of 0 and CWmax. In cases, if the transmission not successful, the CW will be double until it is equal to CWmax while the transmission is successful the backoff timer will decrease to 0. CW process in random backoff time is shown in Figure 3.

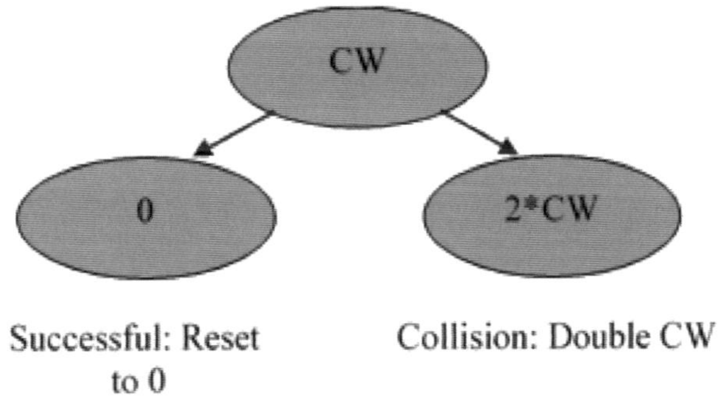

Figure 3. Random Backoff Time

In this section, we analyze the performance of basic CSMA/CA in IEEE 802.11b. The first comparison is made by varying of the number access terminal. From Figure 4, we can infer obvious change in the average delay packets with offered traffic. In other words, the lower number access terminal of performances is better than the highest number access terminal. While considering the impact of the capture effect the performance of average delay packets is better than the without capture effect.

Figure 5 plots the number of access terminal versus offered traffic of basic CSMA/CA. We observe that, the number of access terminal is decreased at an access terminal, transmitted to the access point while the distance between the access point and an access terminal is decreased and signal processing time at access point is better than number of access terminal is increased. Besides that, we considering, the impact of the capture effect the performance of average delay packets is better than the without capture effect.

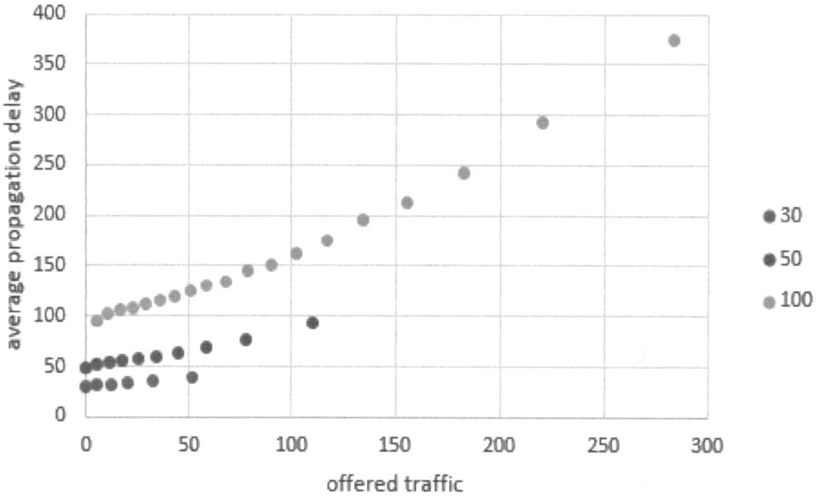

(a)

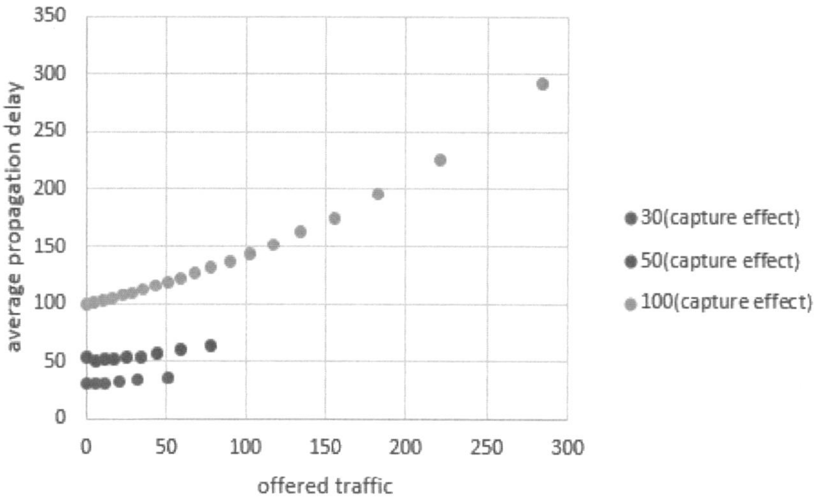

(b)

Figure 4. Average propagation delay versus offered traffic of basic CSMA/CA for varying of the of access terminal (distance = 140 m, capture effect=10 dB) (a) with capture effect and (b) without capture effect

As shown in Figure 6, the third analysis is to discuss the effect on the normalized propagation delay of the offered traffic. The system with capture effect has higher performance than the without capture effect. We observe that, as normalized propagation delay is decreased the system has better performance than the increased normalized propagation delay.

Last analysis is to discuss the effect on the normalized propagation delay on the offered traffic. We observe that, the normalized propagation delay is decreased when the distance between the access point and an access terminal is decreased and signal processing time at access point is better than number of access terminal is increased. On the other hand, the system with capture effect has higher performance than the without capture effect.

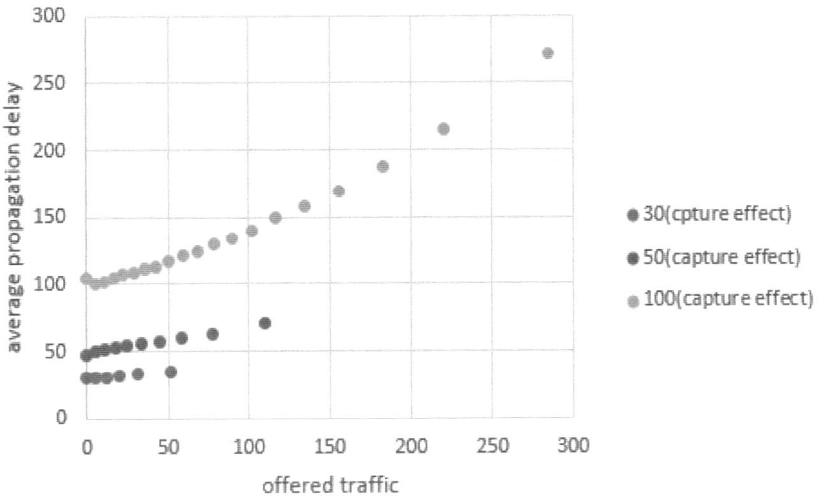

(a)

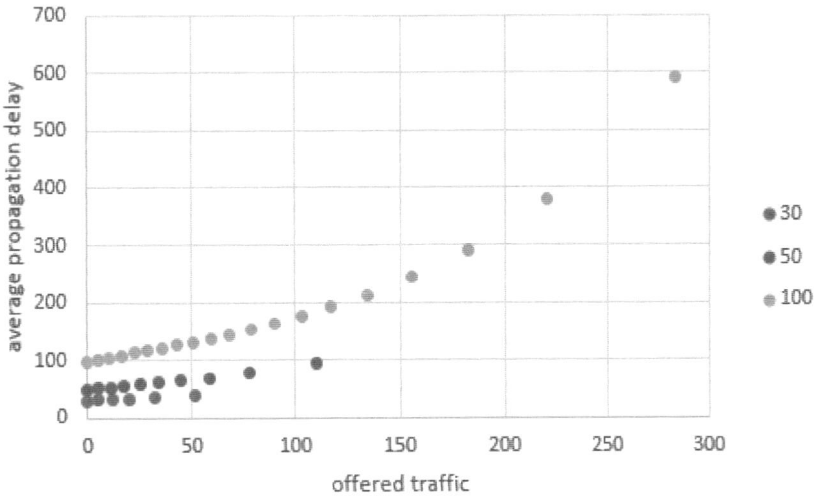

(b)

Figure 5. Average propagation delay versus offered traffic of basic CSMA/CA for varying of the of access terminal (distance = 35 m, capture effect = 10 dB (a) with capture effect and (b) without capture effect

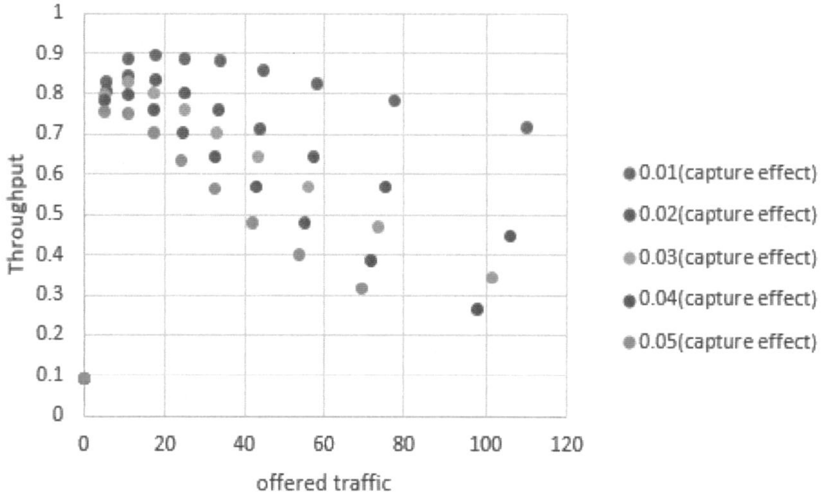

(a)

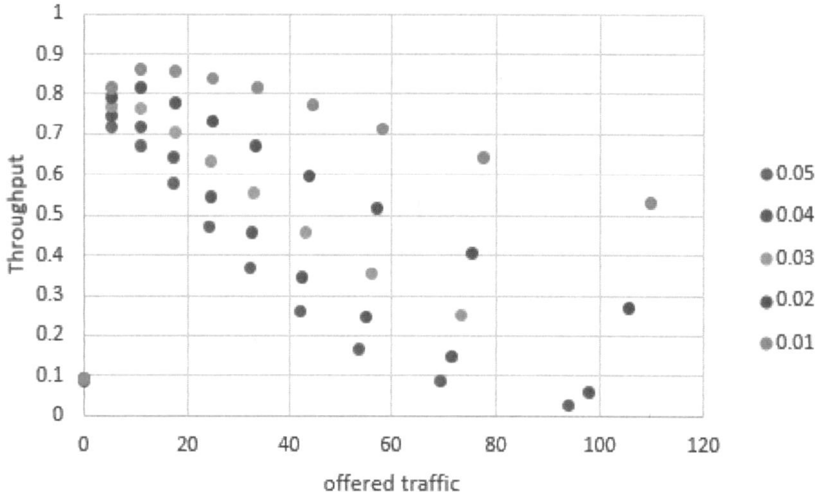

(b)

Figure 6. Throughput versus offered traffic of basic CSMA/CA for varying of the of propagation delay (access terminal = 50, distance = 140 m, capture effect = 10 dB) (a) with capture effect and (b) without capture effect

5. Conclusion

In this paper, different MAC protocol models in IEEE802.11b are used to estimate the network performance by considering the path loss and shadowing for two distance values which are 35 m and 140 m. We can conclude that that shorter distance has higher performance in path loss and shadowing than the longer distance. These results can be used as a useful guide to scientist and engineers before the communication network deployed to transfer captured data to the gateway or control center.

Acknowledgments
The authors would like to acknowledge the support provided by Fakulti Kejuruteraan Elektronik dan Kejuruteraan Komputer (FKEKK), Universiti Teknikal Malaysia Melaka (UTeM). This project is funded by Short Term Grant (Projek Jangka Pendek PJP) in FKEKK, UTeM (PJP/2018/FKEKK/(3B)/S01615). This project is partially funded by Universiti Teknologi Malaysia (UTM) under Vote No. 04G19 and Ministry of Education (MOE) under 4F966 research grants (FRGS).

References

[1] Gummalla. Ajay Chandra Limb.John O, "MEDIUM ACCESS CONTROL PROTOCOLS," IEEE Commun. Surv. Tutorials, vol. 3, no. 2, pp. 2–15, 2000.
[2] A. Mendez-Perez, M. Panduro-Mendoza, R. Aquino-Santos, and M. Munguia-Macario, "Performance Evaluation of Medium Access Control Protocol for Wireless Networks," 2011 IEEE Electron. Robot. Automot. Mech. Conf., pp. 355–359, Nov. 2011.
[3] D. G. Jeong and S. Jeon, "Performance of an Exponential Backoff Scheme for Slotted-ALOHA Protocol in Local Wireless Environment," IEEE Trans. Veh. Technol., vol. 44, no. 3, pp. 470–479, 1995.
[4] W. Ye and J. Heidemann, Medium Access Control in Wireless Sensor Networks. 2004.
[5] P. R. Kumar, "Protocols for Media Access Control and Power Control in Wireless Networks," Proc. 40th IEEE Conf. Decis. Control., vol. 2, no. December, pp. 1935–1940, 2001.
[6] R. Babiker, M. Abdelrahman, A. Babiker, A. Mustafa, and A. A. Osman, "A Comparison between IEEE 802.11 n and ac Standards," IOSR J. Comput. Eng. Ver. III, vol. 17, no. 5, pp. 2278–661, 2015.
[7] IEEE Standards Committee, "Part 11 : Wireless LAN Medium Access Control (MAC) and Physical Layer (PHY) Specifications IEEE Computer Society," vol. 2012, no. March, 2012.
[8] P. Rani, V. Chauhan, S. Kumar, and D. Sharma, "A Review on Wireless Propagation Models," Int. J. Eng. Innov. Technol., vol. 3, no. 11, pp. 256–261, 2014.
[9] A. Woo, K. Whitehouse, F. Jiang, J. Polastre, and D. Culler, "The Shadowing Phenomenon: implications of receiving during a collision," Tech. Rep. UCB//CSD-04-1313, UC Berkeley, 2004.
[10] J. Dai, "The System Performance of Wireless CSMA / CA Protocol," vol. 6, no. 3, pp. 226–234, 2004.
[11] Z. Hadzi-Velkov and B. Spasenovski, "Capture effect with diversity in IEEE 802.11b DCF," Proc. - IEEE Symp. Comput. Commun., pp. 699–704, 2003.
[12] C. Effect, A. Nyandoro, L. Libman, and M. Hassan, "Service Differentiation Using the," vol. 6, no. 8, pp. 2961–2971, 2007.

Performance Analysis of Filtered VHF Signals Captured by Lightning Interferometer System

N. A. I. Azmi[1], M. R. Ahmad[1], M. R. M. Esa[2], M. H. M. Sabri[1], D. Periannan[1], B. Y. Seah[1], S. A. Mohamad[2], G. Lu[3], N. Yusop[4], M. M. Ismail[1], V. Cooray[5], A. A. Alkahtani[6] and M. Z. A. AB. Kadir[6]

[1]Atmospheric and Lightning Research Lab, Centre for Telecommunication Research and Innovation (CeTRI), Fakulti Kejuruteraan Elektronik dan Kejuruteraan Komputer, Universiti Teknikal Malaysia Melaka, Hang Tuah Jaya, 76100 Durian Tunggal, Melaka, Malaysia.
[2]Institute of High Voltage and High Current (IVAT), School of Electrical Engineering, Faculty of Engineering, Universiti Teknologi Malaysia (UTM), 81310 Skudai, Johor Bharu, Malaysia
[3]Institute of Atmospheric Physics, Chinese Academy of Sciences, China.
[4]Space Science Centre (ANGKASA), Institute of Climate Change, Universiti Kebangsaan Malaysia, 43600 Bangi, Selangor, Malaysia
[5]Ångström Laboratory, Division for Electricity, Department of Engineering Sciences, Uppsala University, Box 534, S-75121, Sweden
[6]Universiti Tenaga Nasional (UNITEN), Jalan IKRAM-UNITEN, 43000 Kajang, Selangor.

Email: riduan@utem.edu.my

Abstract. A lightning interferometer is a system that can determine the location of Very High Frequency (VHF) radiation of lightning. Such system has been studied and used for several decades ago. In this project, VHF Interferometer is used to determine the location of VHF radiations emitted by Narrow Bipolar Event (NBE) lightning flash by correlating the signals that received at three antennas. In order to get the better result, cross-correlation algorithm technique is employed to process the signals. This algorithm can determine the time delay between the antennas baselines and then determine the azimuth and elevation angles which are used to locate the location of the NBE flash. This interferometer system also can act as an earlier warning system as it detects the first cloud flash and a forecast can be made before cloud-to-ground flashes occur. Therefore, this interferometer system is very useful for the public citizens in protecting themselves from lightning strikes.

Content from this work may be used under the terms of the Creative Commons Attribution 3.0 licence. Any further distribution of this work must maintain attribution to the author(s) and the title of the work, journal citation and DOI.
Published under licence by IOP Publishing Ltd

1. Introduction

Nowadays, there's a lot of interest in lightning-related research. In particular, studies have been done to understand better about lightning phenomenon including three common types which are Intracloud (IC), Cloud to Ground (CG), and Bolts from the Blue (BLB) [1][2]. The locating system is very useful because it can detect and warn an incoming lightning strike anytime. It gives a lot of benefit to public citizens because they will receive the warning message at the very earliest time when lightning strikes are approaching. It may expose the people near the lightning strike location to danger when they are doing the outdoor activities. A lightning detector was created which is able to detect the lightning strike distance will be useful for the public citizens in protecting themselves from lightning strikes. In the end of the project, we need to analyze the differences between original and filtered signal based on signal processing method. Therefore, in this paper there are four objective which are to design and develop the VHF interferometer with two baselines. A part from that, is to process the analog signal of VHF radiation by sing MATLAB application and produce VHF lightning mapping by using interferometer techniques to determine the direction of arrival of radiation.

Several scopes are listed to ensure the research is conducted within its intended boundary. In this paper, it has two parts; first is experimental work. For the experimental part, the measurement was conducted in wide area such as at Pantai Klebang or in open field in UTeM. The second part is software which is signal processing plays the main role in order to locate the lightning location by using cross correlation technique.

2. Methodology

A. Interferometer Design

The VHF antenna system based on interferometer technique is designed and developed as shown in Figure 1 where the system should be arranged perpendicularly and crossed baselines to each other [4]. This perpendicular baseline interferometer can locate the VHF radiation sources hence VHF radiation source signal received at three capacitive antennas, the signal undergoes radio system that includes bandpass filter 40-80 MHz bandwidth at centered frequency 60 MHz [5]. The signal transfer to LNA in order to amplify the weak signals that received by the antennas. By using LNA, the effect of noise also could be reduced by the gain of an amplifier. After that, the analogue signal will be processed to the digital signal which is transfer the signal into the PicoScope 3000 series via the cable. PicoScope is a tool that can display the voltage waveforms of the signal by using a computer. Then, MATLAB application will take an action to do the signal processing part for plotting the lightning mapping that retrieved from PicoScope. The lightning mapping can be done by using the cross-correlation technique in signal processing that locates the VHF radiation sources. Once the MATLAB coding is run correctly, the resulted of lightning mapping is displayed in both azimuthal and elevational. So that the result obtained can be analyzed and presented.

Communication channel is the path that data takes from transmitter to the destination. Communication channel also known as transmission media or communication media. Communication channel is divided into types which are guided (wired) and unguided (wireless).

In wired communication information is transferred between two or more nodes using wires while in wireless communication information is transferred between two or more nodes that are not connected using electrical conductor. An information is transmitted via open space known as radio waves [8]. Radio waves are generated by radio transmitter and received at the radio receiver. During the transmission from a transmitting antenna to a receiver, path loss occurs due to the propagation medium. The basic propagation loss is shown in Figure 1.

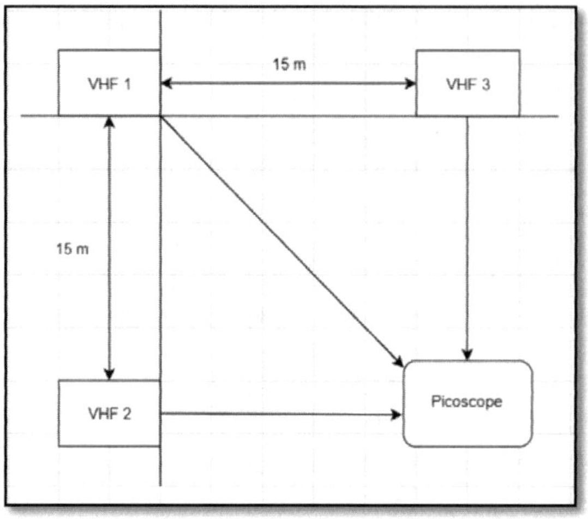

Figure 1. Very High Frequency (VHF) Interferometer System

B. Signal Processing

To process the lightning signal that retrieved from PicoScope, cross-correlation technique is used which estimate time delay. Some equations were involved in order to get the better accuracy of the time delay.

$$d\cos\alpha = c\tau_d = \left(\frac{\Delta_\emptyset}{2\pi}\right)\lambda \quad (1)$$

Based on the equation (1), the value of direction cosine and τ_d are different. So that the equation should be modified as equation (2) and (3).

$$d\cos\alpha = c\tau_{d1} = \left(\frac{\Delta_\emptyset}{2\pi}\right)\lambda \quad (2)$$

$$d\cos\beta = c\tau_{d2} = \left(\frac{\Delta_\emptyset}{2\pi}\right)\lambda \quad (3)$$

To determine azimuth and elevation angle, equations (4) and (5) are needed.

$$\cos\alpha = sin(Az)cos(El) \quad (4)$$

$$\cos\beta = cos(Az)cos(El) \quad (5)$$

From equations (4) and (5) it can be derived to equations (6) and (7).

$$\cos\alpha = \frac{c\tau_{d1}}{d} \quad (6)$$

$$\cos\beta = \frac{c\tau_{d2}}{d} \quad (7)$$

Then $\cos\alpha$ divide by $\cos\beta$ in equations (6) and (7).

$$\frac{\cos\alpha}{\cos\beta} = \frac{\tau_{d1}}{\tau_{d2}} = \frac{\sin Az}{\cos Az} \quad (8)$$

$$\tan Az = \frac{\tau_{d1}}{\tau_{d2}} \quad (9)$$

The azimuth angle then can be calculated by derived from equation (9) as

$$Az = arctan\left(\frac{\tau_{d1}}{\tau_{d2}}\right) \quad (10)$$

Equation (4) is needed to be derived the elevation angle equation.

$$\cos El = \frac{\cos\alpha}{\sin Az}$$

$$\cos El = \frac{\frac{c\tau_{d1}}{d}}{\sin arctan\left(\frac{\tau_{d1}}{\tau_{d2}}\right)} \quad (11)$$

Since,

$$\sin[\arctan(x)] = \frac{x}{\sqrt{1+x^2}}$$

$$\cos El = \frac{c\tau_{d1}\left(\sqrt{1+\left(\frac{\tau_{d1}}{\tau_{d2}}\right)^2}\right)}{d\left(\frac{\tau_{d1}}{\tau_{d2}}\right)}$$

$$\cos El = \frac{c\tau_{d1}\left(\sqrt{\frac{\tau_{d1}{}^2+\tau_{d2}{}^2}{\tau_{d2}{}^2}}\right)}{d\left(\frac{\tau_{d1}}{\tau_{d2}}\right)}$$

$$\cos El = \frac{c\frac{\tau_{d1}}{\tau_{d2}}\left(\sqrt{\tau_{d1}{}^2+\tau_{d2}{}^2}\right)}{d\left(\frac{\tau_{d1}}{\tau_{d2}}\right)}$$

$$\cos El = \frac{c}{d}\sqrt{\tau_{d1}{}^2+\tau_{d2}{}^2} \quad (12)$$

For equation (5),

$$\cos El = \frac{\cos\beta}{\cos Az}$$

$$\cos El = \frac{\frac{c\tau_{d2}}{d}}{\cos arctan\left(\frac{\tau_{d1}}{\tau_{d2}}\right)}$$

Since,

$$\cos[\arctan(x)] = \frac{1}{\sqrt{1+x^2}}$$

$$\cos El = \frac{\frac{c\tau_{d2}}{d}}{1\Big/\sqrt{1+\left(\frac{\tau_{d1}}{\tau_{d2}}\right)^2}}$$

$$\cos El = \frac{c\tau_{d2}\left(\sqrt{1+\left(\frac{\tau_{d1}}{\tau_{d2}}\right)^2}\right)}{d}$$

$$\cos El = \frac{c\tau_{d2}\left(\sqrt{\frac{{\tau_{d1}}^2+{\tau_{d2}}^2}{{\tau_{d2}}^2}}\right)}{d}$$

$$\cos El = \frac{c}{d}\sqrt{{\tau_{d1}}^2+{\tau_{d2}}^2} \qquad (13)$$

The derived elevation equation is then given as:

$$\boldsymbol{El = \arccos\left(\frac{c}{d}\sqrt{{\tau_{d1}}^2+{\tau_{d2}}^2}\right)} \qquad (14)$$

3. Results and analysis

To plot all the samples from the lightning signal, loop function has been used to repeat the same procedure as a single window. The 256 samples were shifting about 32 samples. Because of that, the first window and the next window should be overlapped according to Stock et. al [3]. Figure 2 shows the visualization of the lightning signal which is overlapped and shifting about 32 samples. From the shifting process, the next process is looping. The single window needed to repeat the same procedure in order to mapping all the samples in the same figure. However, the time difference, τ(d) which is for 'tauBC' and 'tauBD' should not exceed the transit time, τ_transit=(d/c) [3]. So that, the value of 'tauBC' and 'tauBD' must fulfill the condition. Figure 3 shown the 'tauBC'for original signal that are not exceed than time transit τ_transit, $5x10^{-8}$ s. Figure 5 shows the lightning mapping of the original signal for the lightning strike. Figure 4 shows the lightning mapping for filtered signal. The pattern of the lightning mapping must be concentrated in a certain area and can be proved in Figure 4 and Figure 5.

Overall, to compare the result between original and filtered signal, the pattern of the filtered signal mapping is almost same as the original signal. However, there has been a little bit different especially for the movement of the lightning. This means the time for every location is a bit different based on the mapping in the figure. The area for the lightning strike should be happening in a certain area. Otherwise, it can be noise, or another lightning flash occurred in the same time.

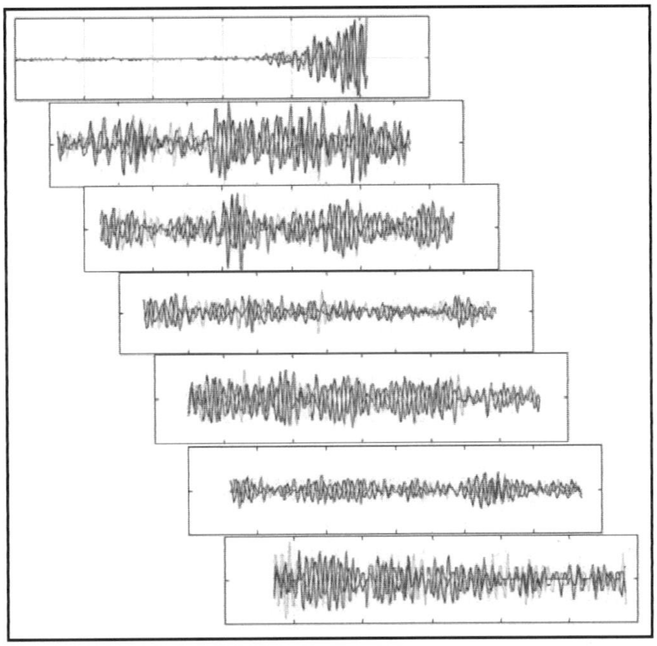

Figure 2. Visualization for Overlapped windows

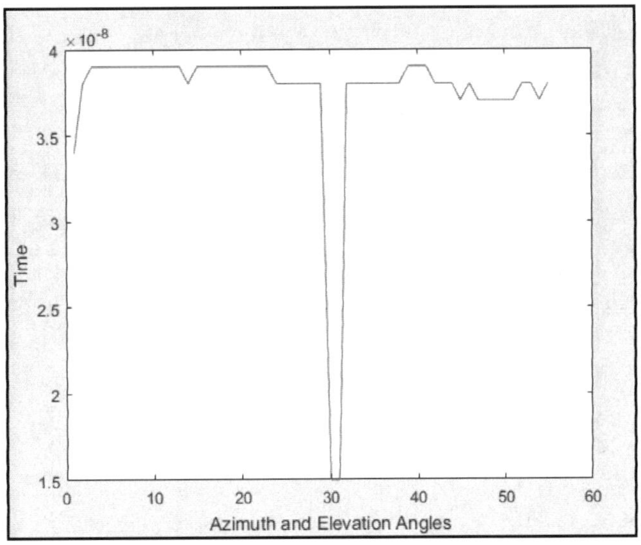

Figure 3. The time difference of 'tauBC' for original signal

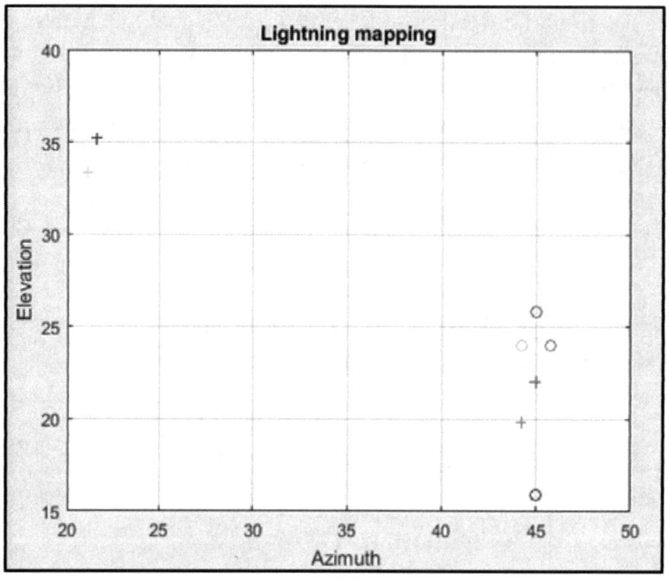

Figure 4. Lightning mapping for filtered signal

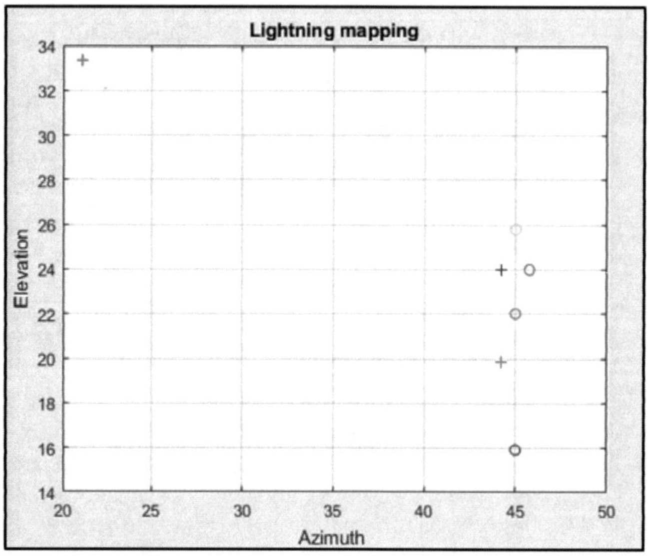

Figure 5. Lightning mapping for original signal

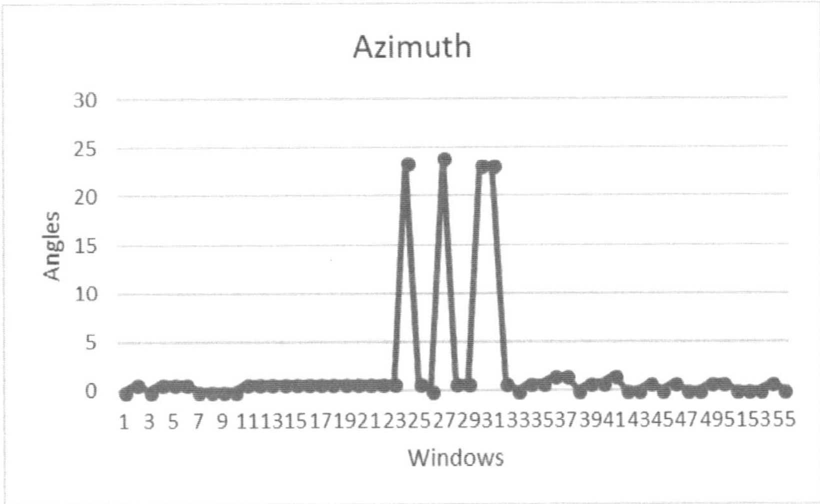

Figure 6. Differences between Original and Filtered Signal for Azimuth Angles

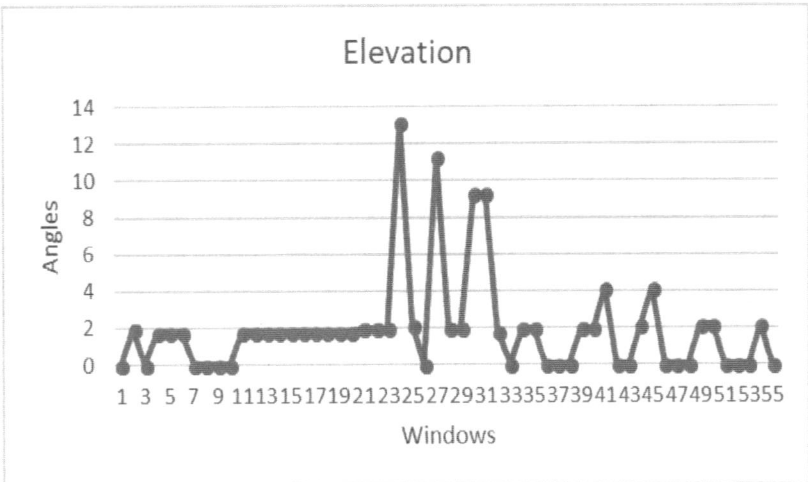

Figure 7. Differences between Original and Filtered Signal for Elevation Angle

4. Conclusion

The purpose of this research is about designing and developing the lightning location system using VHF Interferometer technique. The first objective is to design and develop the VHF Interferometer with two crossed baselines were achieved. Three capacitive antenna was placed in the wide area at Pantai Klebang. These antennas were arranged in perpendicular baseline and successfully received by three antennas as shown in Figure 1.

The second objective is to process the analogue signal of VHF radiation by using MATLAB application. The VHF signal that captured by antennas were undergone bandpass filter and low noise amplifier before transfer to PicoScope through a coaxial cable.

The third objective of this project is to produce VHF lightning mapping by using interferometer technique to determine the time of arrival, τ_d. The cross-correlation technique had been used to correlate the signal that received between two antennas. The result of lightning mapping can present in azimuth and elevation using several equations.

Acknowledgments

The authors would like to acknowledge the support provided by Fakulti Kejuruteraan Elektronik dan Kejuruteraan Komputer (FKEKK), Universiti Teknikal Malaysia Melaka (UTeM). This project is funded by Short Term Grant (Projek Jangka Pendek PJP) in FKEKK, UTeM (PJP/2018/FKEKK/(3B)/S01615). This project is partially funded by Universiti Teknologi Malaysia (UTM) under Vote No. 04G19 and Ministry of Education (MOE) under 4F966 research grants (FRGS).

References

[1] Proctor, D., VHF radio pictures of cloud flashes. Journal of Geophysical Research. 1981, 86(C5), 4041-4071. 1,

[2] Rison, W., R. Thomas, P. Krehbiel, T. Hamlin, and J. Harlin. A GPS based three dimensional lightning mapping system, Geophysical research letters. 1999, 26(23), 3573-3576. 1, 6.

[3] Stock, M., M. Akita, P. Krehbiel, W. Rison, H. Edens, Z. Kawasaki, and M. Stanley. Continuous broadband digital interferometry of lightning using a generalized cross-correlation algorithm. Journal of Geophysical Research: Atmospheres. 2014, 119 (6), 3134-3165. 15, 26, 39, 40, 64, 65, 103,

[4] Hayenga, C., and J. Warwick. Two-dimensional interferometric positions of VHF lightning sources. Journal of Geophysical Research. 1981, 86 (C8), 7451-7462. 3, 4, 39

[5] Warwick, J. W., C. O. Hayenga, and J. W. Broshanan (1979), Interferometric directions of lightning sources at 34 MHz, Journal of Geographic Research: Oceans (1978-2012), 84 (C5), 2457-2468. 1,2,3.

Performance Analysis of Flame Retardant 4 Copper Plate Antenna for Lightning Remote Sensing

D. Periannan[1], S. A. Mohamad[2], M. R. Ahmad[1], M. R. M. Esa[2], M. H. M. Sabri[1], B. Y. Seah[1], G. Lu[3], N. Yusop[4], M. M. Ismail[1], Z. Abdul-Malek[2], V. Cooray[5], A. A. Alkahtani[6] and M. Z. A. AB. Kadir[6]

[1]Atmospheric and Lightning Research Lab, Centre for Telecommunication Research and Innovation (CeTRI), Fakulti Kejuruteraan Elektronik dan Kejuruteraan Komputer, Universiti Teknikal Malaysia Melaka, Hang Tuah Jaya, 76100 Durian Tunggal, Melaka, Malaysia.
[2]Institute of High Voltage and High Current (IVAT), School of Electrical Engineering, Faculty of Engineering, Universiti Teknologi Malaysia (UTM), 81310 Skudai, Johor Bharu, Malaysia
[3]Institute of Atmospheric Physics, Chinese Academy of Sciences, China.
[4]Space Science Centre (ANGKASA), Institute of Climate Change, Universiti Kebangsaan Malaysia, 43600 Bangi, Selangor, Malaysia
[5]Ångström Laboratory, Division for Electricity, Department of Engineering Sciences, Uppsala University, Box 534, S-75121, Sweden
[6]Universiti Tenaga Nasional (UNITEN), Jalan IKRAM-UNITEN, 43000 Kajang, Selangor.

Email: riduan@utem.edu.my

Abstract. Lightning remote sensing has been used widely to measure the electromagnetic (EM) fields in various frequencies bands. Circular metal plate capacitive antenna made of iron or aluminium have been used widely for the front-end design of the remote sensing system. The circular metal plates are arranged in parallel and separated by insulators made of Teflon. However, three common problems encountered with the existing setting. First, both metals rust easily and could affect the reading of the EM fields. Second, the insulation Teflon separating the metal plates could become weak conductor during heavy rain and thus distort the EM fields reading. Third, the use of Teflon in between the parallel plates changes the overall permittivity value. In this paper, we propose a cheaper and lighter alternative to iron and aluminium plates by changing to FR4 copper plate. Rectangular FR4 copper plate capacitive antennas with 2 sizes, A4 and A3 have been designed and constructed. The rectangular A3 FR4 copper plate has been found to have similar impedance and capacitance values to the iron-based circular metal plate. Analysis of wave shape and peak amplitude ratio showed similar performance between both antennas. Therefore, we conclude that rectangular A3 FR4 copper plate antenna can be used as replacement for the existing iron- or aluminium-based circular metal plates antenna.

Content from this work may be used under the terms of the Creative Commons Attribution 3.0 licence. Any further distribution of this work must maintain attribution to the author(s) and the title of the work, journal citation and DOI.
Published under licence by IOP Publishing Ltd

IOP Conf. Series: Earth and Environmental Science **228** (2019) 012006 doi:10.1088/1755-1315/228/1/012006

1. Introduction

To measure a lightning event, the measurement of electric field can be done by using a single plate or parallel plate capacitive antenna. Gauss's law is one of the primary laws that is applied during the calibration process, where it links between the charge in a region or a surface to the electric field. In Coulomb's law, it focused on the force of two charges or any number of charges where the concept of electric field is defined by force per unit charge for each point in a region of charges [1]. For a single plate capacitive antenna, the electric field is assumed as uniform and the antenna is considered electrically small. No metallic object should be place near the system to avoid any disturbance during the measurement of electric field [2]. When the antenna is in no load condition, the electric field near the antenna matched to the value that would exist in the absence on antenna. The potential difference between the ground and the antenna can then be rule by the equation of $\boldsymbol{Vg = En.d}$ [3]. In this equation, the voltage between the metallic plate and the ground is directly proportional to the electric field normal to the plate and its height respect to the ground.

Lightning is measured based on their electric field intensity and one of the crucial aspects of capturing the information is the operation of the measuring system. This system includes the geometry of the antenna, electronic circuits of the buffer circuits and the calibration of the system. The measuring system consists of a flat metallic plate above ground level to measure the lightning electric field intensity. There are two parameters that were observed for this measuring system which are the size of the antenna and the capacitive value at the electronic buffer circuit. Changing both parameters affects the potential readings for the Vg and Vm values. These parameters were simulated in the Matlab by using two main equations that would calculate the outcome of the potential at the antenna and the output. Based on the results obtained, the best design could be determined.

Circular metal plate capacitive antenna made of iron or aluminium have been used widely for the front-end design of the remote sensing system. The circular metal plates are arranged in parallel and separated by insulators made of Teflon. However, three common problems encountered with the existing setting. First, both metals rust easily and could affect the reading of the EM fields. Second, the insulation Teflon separating the metal plates could become weak conductor during heavy rain and thus distort the EM fields reading. Third, the use of Teflon in between the parallel plates changes the overall permittivity value. In this paper, we propose a cheaper and lighter alternative to iron and aluminium plates by changing to FR4 copper plate. Rectangular FR4 copper plate capacitive antennas with 2 sizes, A4 and A3 have been designed and constructed.

2. Methodology

A parallel plate antenna with air gap has been designed in Computer Simulation Technology (CST) simulator by using static and low frequency solver as shown in Figure 1. The Flame Retardant 4 (FR4) substrate was constructed using brick model with the size of 420 mm x 297 mm x 1.6 mm. The material used was Perfect Electric Conductor (PEC). Then, another brick model with the size of 380 mm x 297 mm x 0.035 mm was constructed and the material used was annealed copper. Later, another substrate and copper with the same size as previous ones were constructed. Both the copper plates were faced inside while the two substrates were faced outside just like mirrored. The spacing (air gap) between the copper plates was approximately 30 mm.

Next, by using the static and low frequency components, electric field was applied between the plates by define the total charges for both copper plates (+Q for top plate and –Q for bottom plate) and been simulated. The results for the simulation of the parallel plate are shown in Figures 2 and 3.

IOP Conf. Series: Earth and Environmental Science **228** (2019) 012006 doi:10.1088/1755-1315/228/1/012006

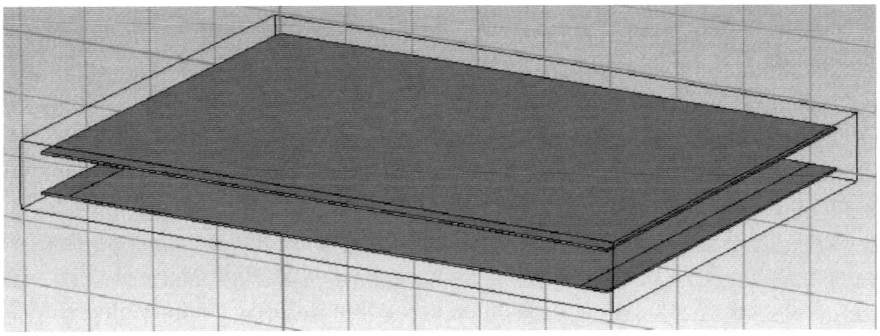

Figure 1. Model of a parallel plate with thin copper plates facing each other and separated by 3 cm air gap

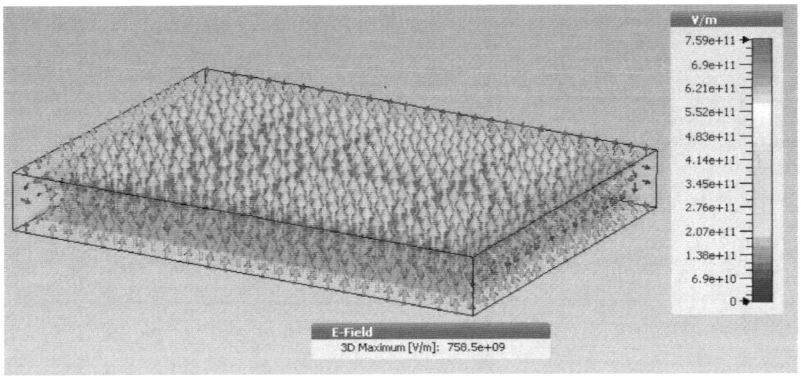

Figure 2. Plot of 3D Electric field between the copper plates. It is clear that the electric field is almost uniform between the plates (green arrows) except at the edges due to fringing effect

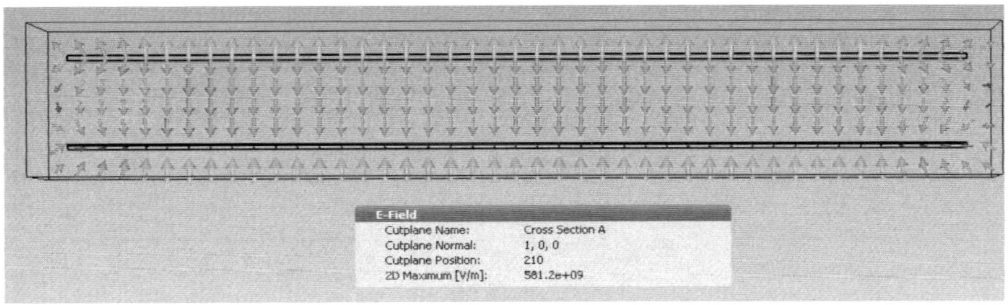

Figure 3. Plot of 2D Electric field between the copper plates where the direction of the vertical field is downward as shown by the green arrows. This is because the top plate was defined to have positive surface charge density (+Q) while the bottom plate has negative surface charge density (-Q). The fringing effect could be seen clearly at the both edges of the plot

Figures 2 and 3 show that vertical electric field was generated between the parallel plates directed downward. The electric fields flow from positively-plate (top plate) to negatively-plate (bottom plate). The simulation of parallel plate capacitive antenna is considered a success due to the uniform vertical electric field generated between the plates which perfectly agree with Gauss's Law for Electrostatic.

As can be seen from both the 2D and 3D plots, the direction of the electric fields is uniform but at the edge of the antenna, notice that the fields are not directed vertically downward between the plates. The fields curving as the fields flow from the top plate to the bottom plate. This phenomenon is known as fringing effect that happens at the edges of an antenna and this problem is very severe for high frequency design. In this thesis, we design the antenna for low frequency application below than 3 MHz frequency band. Therefore, fringing effect can be neglected completely which will not affect the measurement campaign.

3. Results and analysis

3.1 Antenna performance under real lightning measurement

Figure 4 shows the variation of the electric fields between metal plate (iron) and FR4 copper plate antenna with A3 size. The A3 size was chosen because of comparable dimension when compared to the dimension of the metal plate antenna. This data has been taken on November 16th 2015 to compare the performance between metal plate antenna and FR4 A3 antenna. 100 selected data have been chosen to do comparison study between both antennas. The selected type of waveform chosen is Negative Cloud-to-Ground (–CG) flash and Positive Narrow Bipolar Event (+NBE). These 2 types of waveforms frequently detected by Picoscope. The peak values of the electric field were obtained from the Picoscope record (recorded as voltage value) at Channel A (metal plate) and Channel B (FR4 copper plate with A3 size). For analysis purpose, the ratio of the peak values of Channel A to Channel B is obtained.

As can be seen from Figure 4, the ratio values vary in a range between 0.8 to 1.06. The mean value is calculated to be around 0.935 which tells us that the peak amplitudes of the FR4 copper plate antenna are almost the same as the peak amplitudes of the metal plate antenna. This result is expected because the size of the copper plate antenna is not exactly the same as metal plate antenna. Thus, small difference is really expected. The variance and standard deviation are calculated to be around 0.004 and 0.063, respectively.

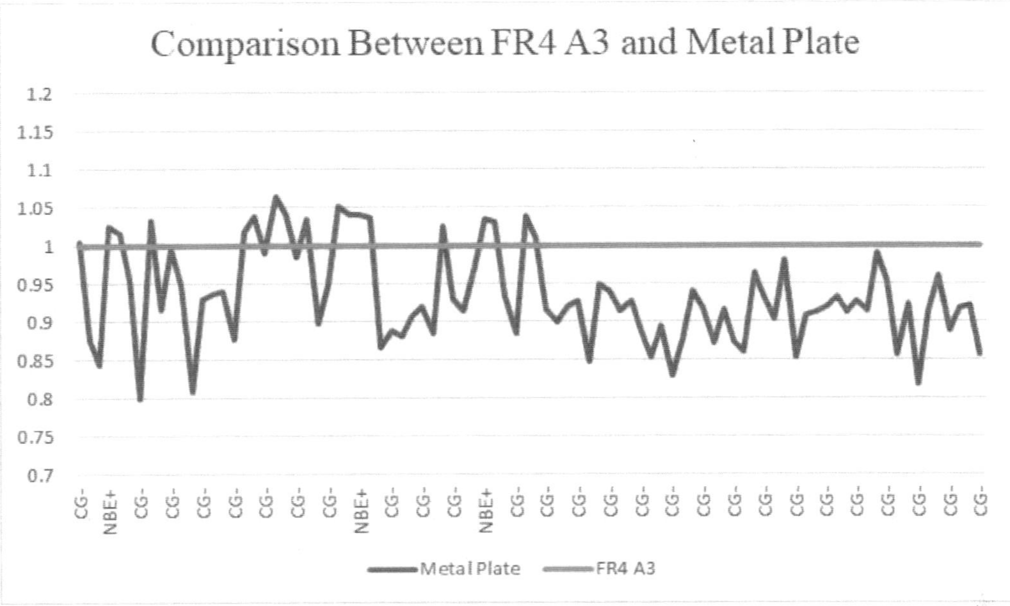

Figure 4. Plot shows the performance comparison between metal plate antenna and FR4 copper plate antenna for both –CG and +NBE flashes. The ratios are determined based on peak electric field values of metal plate over the values of FR4 antenna

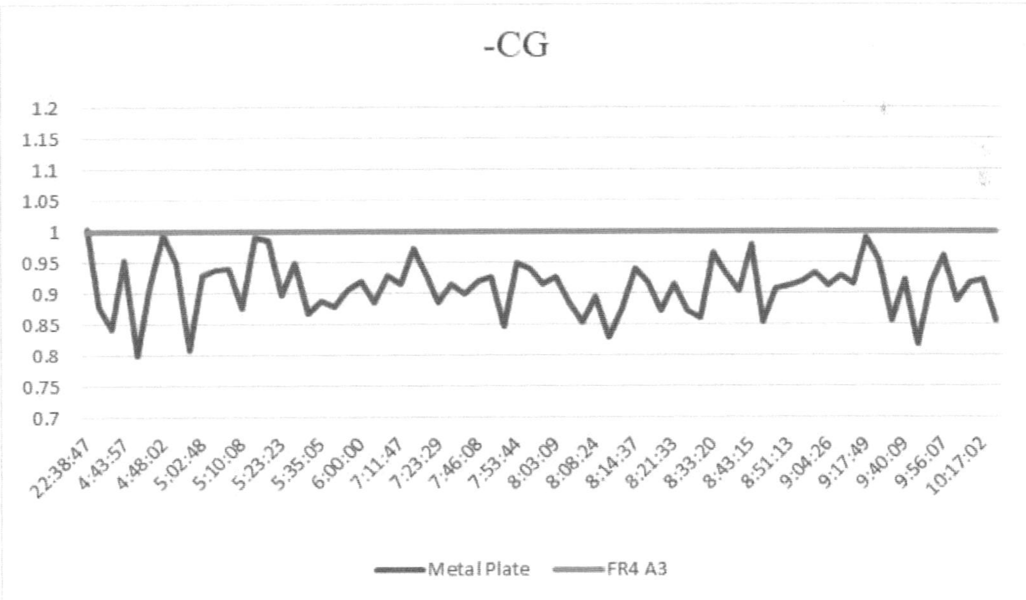

Figure 5. Plot shows the performance comparison between metal plate antenna and FR4 copper plate antenna for –CG flashes. The ratios are determined based on peak electric field values of metal plate over the values of FR4 antenna

Figure 5 shows the variation of the electric fields between metal plate (iron) and FR4 copper plate antenna with A3 size for only –CG flashes. By comparing the performance of the captured –CG flashes, we can see that the FR4 copper plate antenna performance is comparable to the metal plate antenna. The mean value is calculated to be around 0.911 which tells us that the peak amplitudes of the FR4 copper plate antenna are almost the same as the peak amplitudes of the metal plate antenna. This result is expected because the size of the copper plate antenna is not exactly the same as metal plate antenna. Thus, small difference is really expected. The variance and standard deviation are calculated to be around 0.002 and 0.044, respectively.

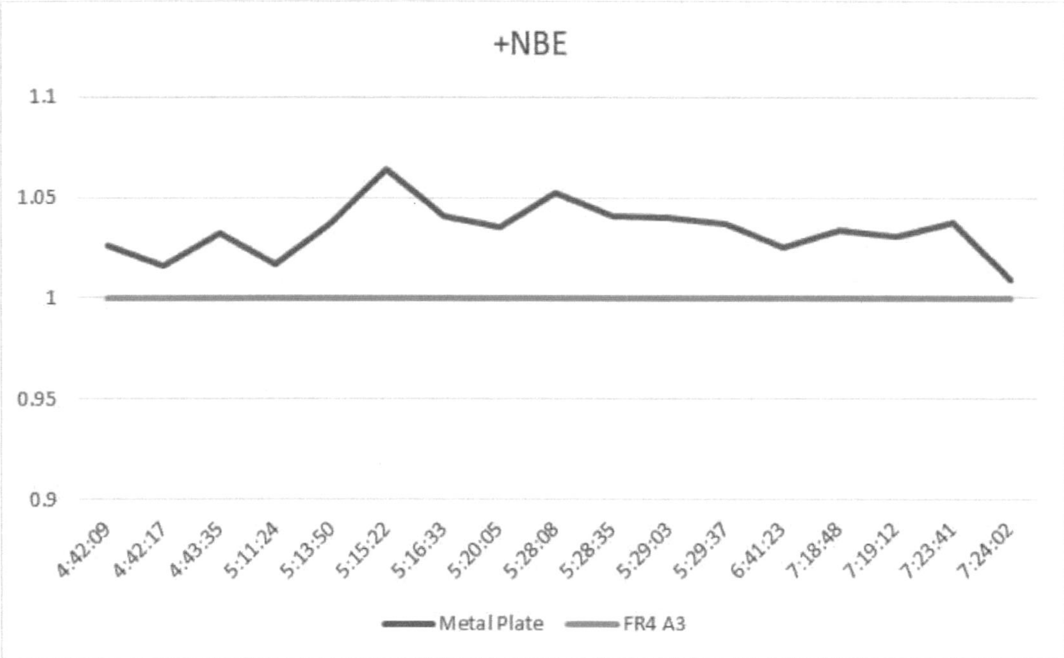

Figure 6. Plot shows the performance comparison between metal plate antenna and FR4 copper plate antenna for +NBE flashes. The ratios are determined based on peak electric field values of metal plate over the values of FR4 antenna

Figure 6 shows the variation of the electric fields between metal plate (iron) and FR4 copper plate antenna with A3 size for only +NBE flashes. By comparing the performance of the captured +NBE flashes, we can see that the FR4 copper plate antenna performance is comparable to the metal plate antenna. The mean value is calculated to be around 1.034 which tells us that the peak amplitudes of the FR4 copper plate antenna are almost the same as the peak amplitudes of the metal plate antenna. This result is expected because the size of the copper plate antenna is not exactly the same as metal plate antenna. Thus, small difference is really expected. The variance and standard deviation are calculated to be around 0.0002 and 0.013, respectively.

3.2 Analysis of captured waveforms

3.2.1 Capture waveforms by using both metal and copper plate antennas

Figures 7 and 8 show the captured waveform of –CG and +NBE flashes, respectively. We observed that the amplitudes of the electric fields of the metal plate antenna were slightly higher than the amplitudes of the electric fields of the FR4 copper plate antenna. This case is true for –CG flashes while for +NBE flashes the case is contrary. This result is expected due to the area of the metal plate antenna is slightly bigger compared to FR4 copper plate size antenna. Moreover, the shape of the metal plate antenna is circular while the FR4 copper plate is rectangular. However, the most important factor here is the shape of the waveforms is identical which means that the frequency components of both antenna systems are preserved.

In summary, the difference between metal plate and copper plate antenna systems is not that significant and we decided to use FR4 copper plate antenna system to replace the old metal plate antenna. The main reason is that the metal plate antenna is much costlier compared to FR4 copper plate antenna. The FR4 copper plate antenna can be purchased easily at any electronic shops. Moreover, FR4 copper plate antenna is much lighter and mobile compared to metal plate that is heavy and easily to corrode.

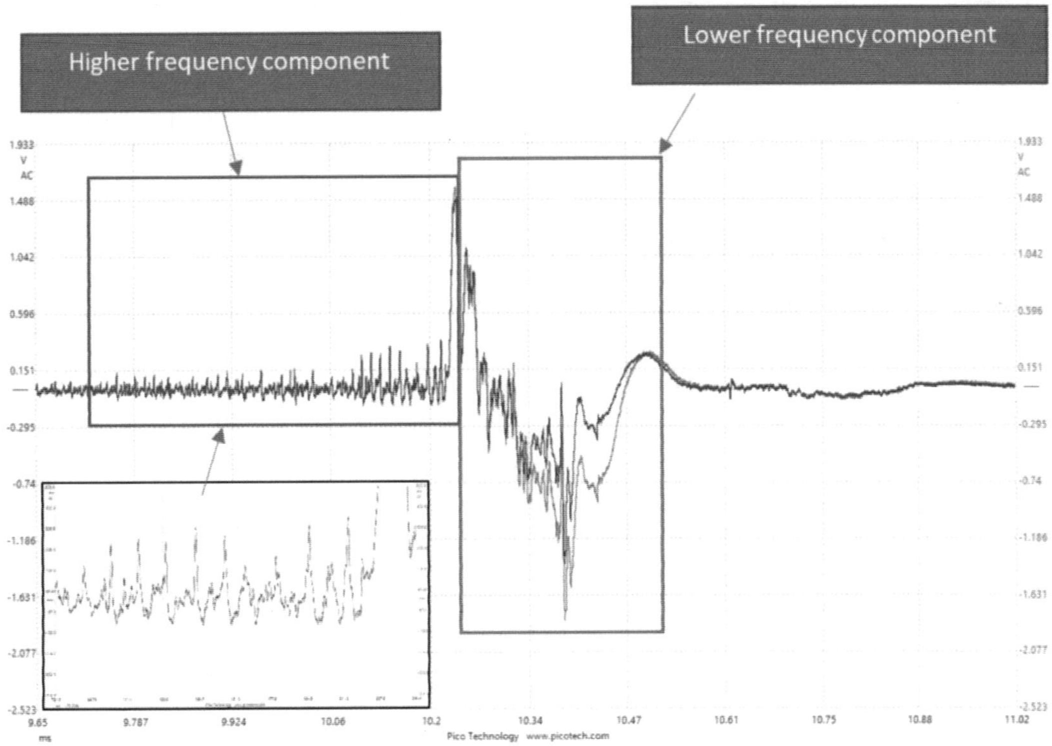

Figure 7. An example of captured waveform of –CG flash that consists of stepped leader and return stroke processes. The blue plot is signal captured by metal plate antenna system and the red plot is signal captured by FR4 copper plate antenna system

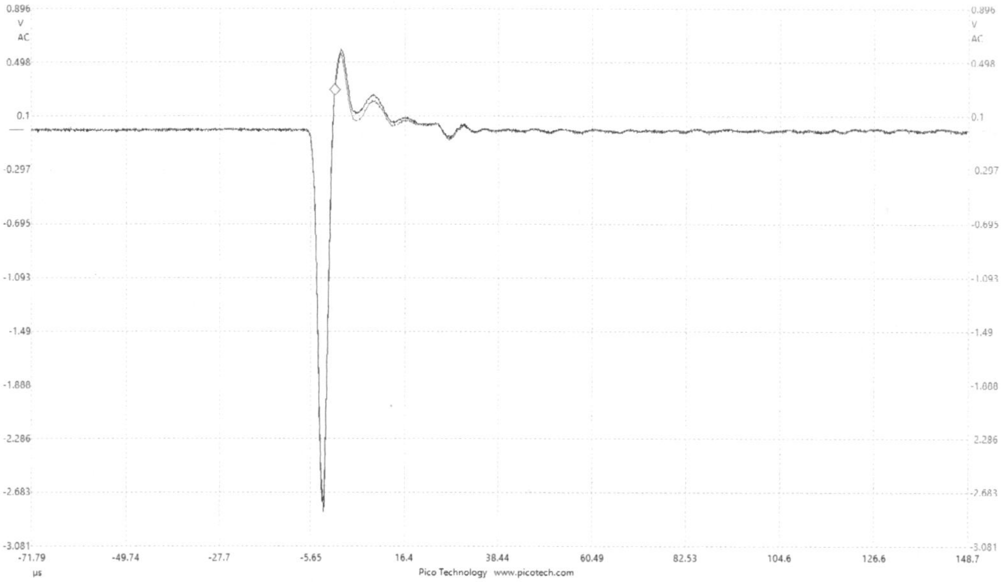

Figure 8. An example of captured waveform of +NBE flash. The blue plot is signal captured by metal plate antenna system and the red plot is signal captured by FR4 copper plate antenna system

3.2.2 Captured waveforms beyond reversal distance by using FR4 copper plate antenna

Figures 9 and 10 show captured waveforms of –CG and IC flashes beyond reversal distance (typically > 7 km [4,5] by using fast field system (top blue waveform) and slow field system (red bottom waveform) [7]. Fast field system captured mainly radiation component of the electric field emitted by the lightning while slow field system captured mainly electrostatic and inductive components of the electric field [6].

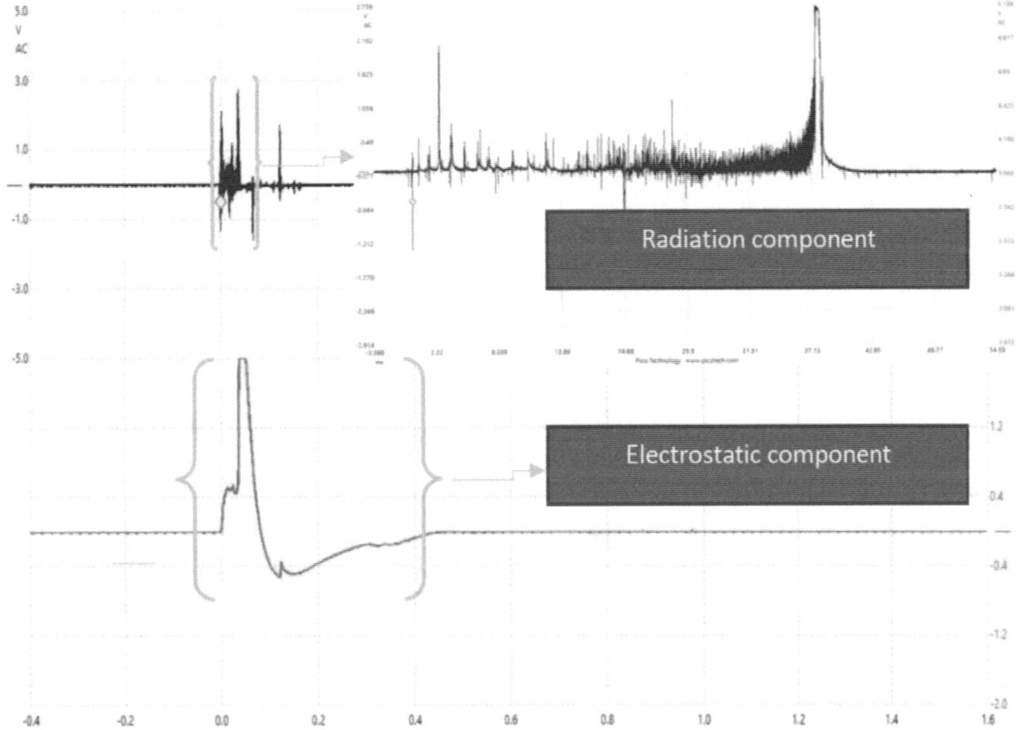

Figure 9. An example of –CG flash captured with radiation component (top blue plot) and electrostatic component (bottom red plot)

As can be seen from the fast field waveform (blue plot) in Figure 9, the captured –CG flash was initiated by preliminary breakdown and followed by step leaders and return stroke. These processes are essential part of a –CG flash [6]. Negative CG flash literally means that negative charges are brought down to the ground. When the negative charges brought down to the ground, the polarity of the emitted electric field is positive (atmospheric electricity sign convention) due to fact that the electric current flows upward [6]. In other word, the polarity of the emitted electric field pulses is determined based on the direction of the current. Therefore, the polarity of the initial part of the slow field record (red plot) should be positive. This is clear evidence that FR4 copper plate antenna could detect lightning waveforms faithfully.

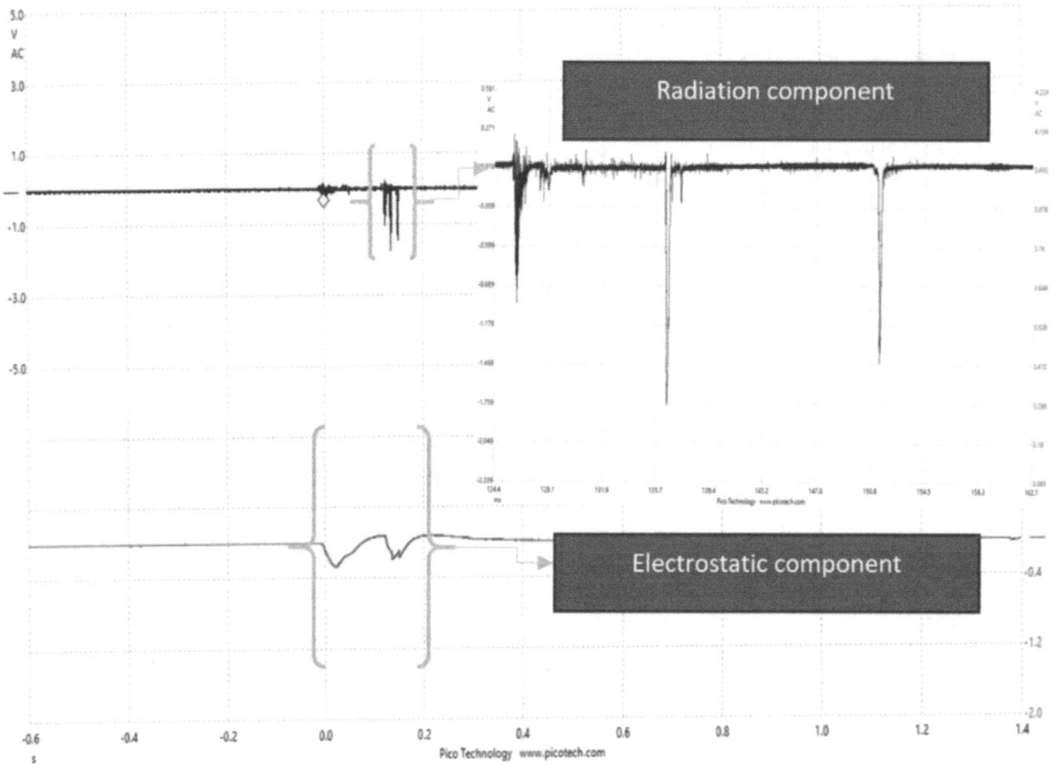

Figure 10. An example of IC flash captured with radiation component (top blue plot) and electrostatic component (bottom red plot)

As can be seen from the fast field waveform (blue plot) in Figure 10, the captured IC flash was initiated by preliminary breakdown and followed by regular bipolar pulses. These processes are essential part of an IC flash [6]. The IC flash literally means that negative charges are brought up to bridge between main negative charge and main positive charge inside thunderstorm. When the negative charges brought up, the polarity of the emitted electric field is negative (atmospheric electricity sign convention) due to fact that the electric current flows downward [6]. In other word, the polarity of the emitted electric field pulses is determined based on the direction of the current. Therefore, the polarity of the initial part of the slow field record (red plot) should be negative. This is clear evidence that FR4 copper plate antenna could detect lightning waveforms faithfully.

3.2.3 Captured waveforms within reversal distance by using FR4 copper plate antenna

Figures 11 and 12 show captured waveforms of –CG and IC flashes within reversal distance (typically ≤ 7 km [4,5] by using fast field system (top blue waveform) and slow field system (red bottom waveform). Within reversal distance, fast field system captured both radiation and electrostatic components of the electric field emitted by the lightning while slow field system captured mainly electrostatic and inductive components of the electric field [6].

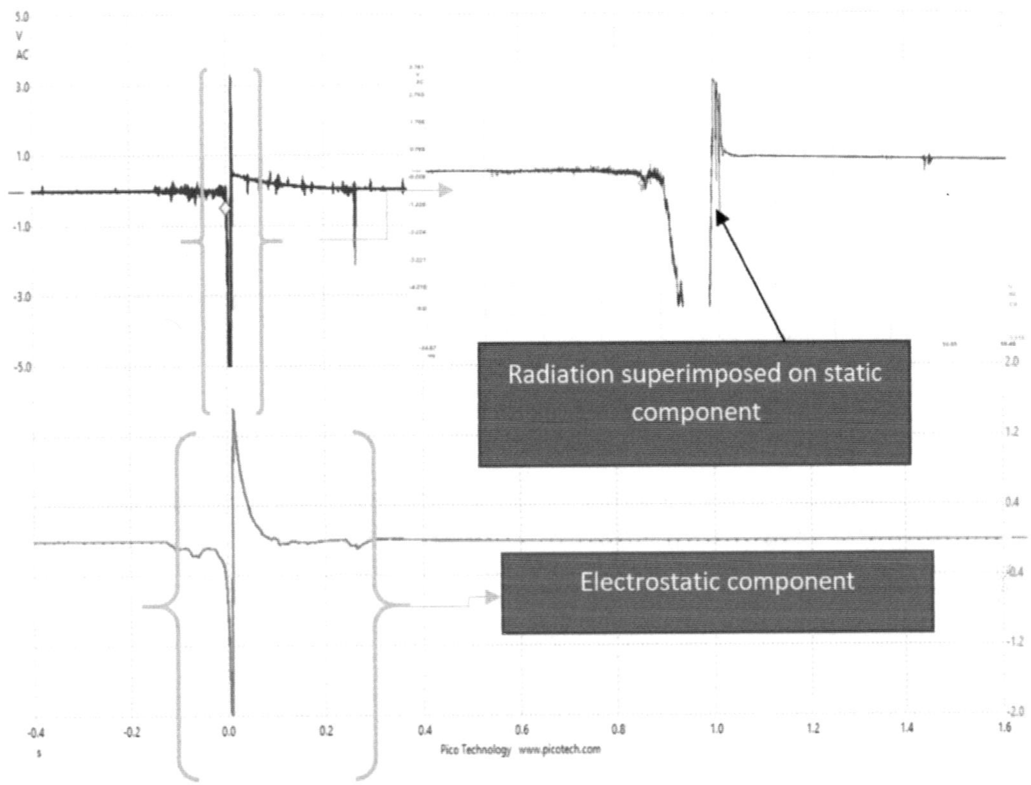

Figure 11. An example of –CG flash captured with radiation component superimposed on static component (top blue plot) and electrostatic component (bottom red plot)

As can be seen from the fast field waveform (blue plot) in Figure 11, the captured –CG flash was initiated by preliminary breakdown and followed by step leaders and return stroke superimposed on static field. These processes are essential part of a –CG flash [6]. Within reversal distance, the polarity of the emitted electric field is negative (atmospheric electricity sign convention) due to fact that the dominant field at this distance is electrostatic field even though the electric current flows upward [6]. Therefore, the polarity of the initial part of the slow field record (red plot) should be negative. This is clear evidence that FR4 copper plate antenna could detect lightning waveforms faithfully.

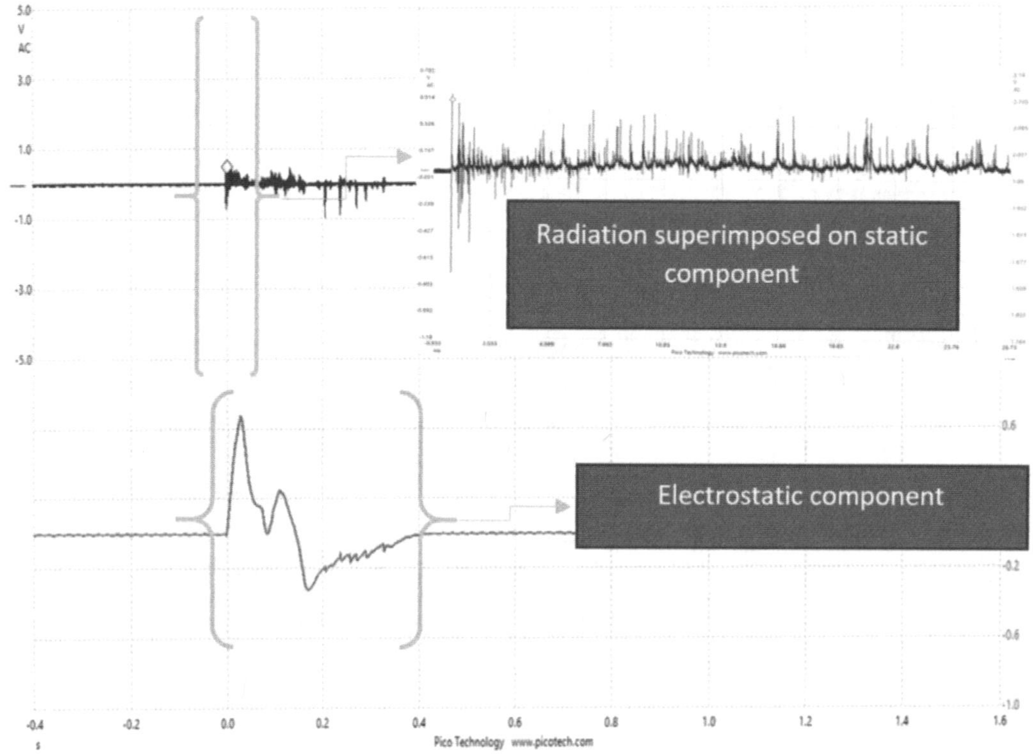

Figure 12. An example of IC flash captured with radiation component superimposed on static component (top blue plot) and electrostatic component (bottom red plot)

As can be seen from the fast field waveform (blue plot) in Figure 12, the captured IC flash was initiated by preliminary breakdown and followed by regular bipolar pulses. These processes are essential part of an IC flash [6]. Within reversal distance, the polarity of the emitted electric field is positive (atmospheric electricity sign convention) due to fact that the dominant field at this distance is electrostatic field even though the electric current flows upward [6]. Therefore, the polarity of the initial part of the slow field record (red plot) should be positive. This is clear evidence that FR4 copper plate antenna could detect lightning waveforms faithfully.

4. Conclusion

The aims of this paper is to design, simulate and fabricate FR4 A3 size antenna for wideband antenna system. The system has been simulated and fabricated successfully and had been deployed for real time measurement purpose. Therefore, the first aim has been achieved successfully. While, the second objective is to do comparison studies between metal parallel plate antenna and FR4 A3 size antenna. 100 lightning samples were captured and analysed with fast field antenna system for both antennas to evaluate the performance. The FR4 A3 size antenna is found to be suitable to replace the existing metal parallel plate antenna because it is light and easy to deploy at any places. So, the second objective has been achieved successfully.

Acknowledgments

The authors would like to acknowledge the support provided by Fakulti Kejuruteraan Elektronik dan Kejuruteraan Komputer (FKEKK), Universiti Teknikal Malaysia Melaka (UTeM). This project is funded by Short Term Grant (Projek Jangka Pendek PJP) in FKEKK, UTeM (PJP/2018/FKEKK/(3B)/S01615). This project is partially funded by Universiti Teknologi Malaysia (UTM) under Vote No. 04G19 and Ministry of Education (MOE) under 4F966 research grants (FRGS).

References

[1] V. Cooray, “An Introduction to Lightning”, Springer 2015.
[2] H. Perez, "Study on the interaction of lightning electromagnetic fields with low voltage power installations", Lic, Thesis, Uppsala University, 1993.
[3] Ace, Version 2.57, ABB common platform for field analysis and simulations, ABB Corporate Research Center, Vasteras, Sweden, 1996.
[4] Marshall, T., Stolzenburg, M., Karunarathna, N., & Karunarathne, S. (2014). Electromagnetic activity before initial breakdown pulses of lightning. Journal of Geophysical Research: Atmospheres, 119(22).
[5] R. Chapman, T. Marshall, S. Karunarathne, and M. Stolzenburg, “Initial electric field changes of lightning flashes in two thunderstorms,” J. Geophys. Res. Atmos., vol. 122, no. 7, pp. 3718–3732, 2017.
[6] Rakov, V.A. and Uman, M.A, 2003. Lightning: physics and effects. Cambridge University Press.
[7] Galvan, A. and Fernando, M., 2000. Operative characteristics of a parallel-plate antenna to measure vertical electric fields from lightning fields from lightning flashes. Report UURIE, pp.285-00.

WiFi Temporal Coverage: Analysis of Socio-Economics Influences in Malaysia

N. A. Wahab[1], M. R. Ahmad[1], A. A. M. Isa[1], R. F. Malik[2] and M. R. M. Esa[3]

[1]Broadband and Networking Research Group (BBNET), Centre for Telecommunication Research and Innovation (CeTRI), Fakulti Kejuruteraan Elektronik dan Kejuruteraan Komputer, Universiti Teknikal Malaysia Melaka, Hang Tuah Jaya, 76100 Durian Tunggal, Melaka, Malaysia.
[2]Faculty of Computer Science, Universitas Sriwijaya (UNSRI), Inderalaya, Sumatera Selatan, Indonesia
[3]Institute of High Voltage and High Current (IVAT), School of Electrical Engineering, Faculty of Engineering, Universiti Teknologi Malaysia (UTM), 81310 Skudai, Johor Bharu, Malaysia

Email: nazdiana.aw@gmail.com; riduan@utem.edu.my; rezafm@unsri.ac.id

Abstract. According to Internet Survey 2017, it was found that 89.4 percent of Internet users in Malaysia used a smartphone as a medium to access the internet. The proliferation of mobile devices and bandwidth hungry application, has resulted in mobile network provider (MNO) struggles to keep up with the ever-increasing demand for higher network capacity and wider network coverage. WiFi offloading seems the most viable solution due to the abundance and readily available infrastructure. Diverted traffic to the WiFi network may possibly reduce cellular network traffic, hence cellular network provider may be able to accommodate traffic growth at a lower cost. By analysing and understanding the user's offloading behaviour may provide guidance in price and cost restructuring strategies or in network planning. This paper investigates temporal coverage of mobile user's WiFi utilization and socio-economics influences through an extensive measurement study. The measurement study is carried out for 18 days and collected real data traces of user WiFi utilization on Android smartphone. The observation analysed user's WiFi temporal coverage where participants are categorized into two different socio economics group. The finding reveals low WiFi temporal coverage for both groups of smartphones users. Group 1 spends more time using WiFi compared to Group 2, while both groups shows higher temporal coverage during active hour. Findings of this research may potentially serve as insights to the relevant stakeholder or regulatory body, as well as a barometer to carry out enhancement measures or any potential joint venture between cellular network provider and ISP.

Content from this work may be used under the terms of the Creative Commons Attribution 3.0 licence. Any further distribution of this work must maintain attribution to the author(s) and the title of the work, journal citation and DOI.
Published under licence by IOP Publishing Ltd

1. Introduction

From the past 10 years, Malaysia has experienced tremendous growth in mobile broadband subscriptions to 30.6 million in the first quarter of 2017, from only 23.35 million in 2007 [1, 2]. MNO has continuously upgrading and expanding their infrastructure to extend network coverage as well as to improve network capacity. However, mobile broadband coverage is not universal. 4G and 3G coverage usually only available at a strategic location such as metropolitan areas, offices and industrial area or location with high population. Most of the places either still experiencing slow mobile broadband service such as EDGE or does not have mobile broadband coverage at all.

In addition, some of the places with good mobile broadband coverage experiencing problems like a blind spot. For indoor building, there are still issue on coverage even though indoor base station or indoor mobile broadband base station (MBB) has been installed inside the building. Mobile users are also experiencing slow data speed during peak hour due to network congestion, particularly at high traffic areas such as at a crowded airport or football stadium.

[3] highlighted one of the most frequent consumer complaints are related to service speed. The signal fluctuated between 3G signal and EDGE signal even though the area is within 3G coverage. The unstable connection to the end user is caused by the fluctuations of the receiving signals.

Therefore, consumers experienced slow speed during data connection activities. The issues mentioned above pointed out some of the limitations of mobile data service provided by the cellular networks. The issues mainly are caused by the limited network capacity or also known as bandwidth scarcity. The problems are further exacerbated by the proliferation of mobile devices and rich multimedia content, video streaming and cloud services that require reliable and high bandwidth capacity.

According to Internet Survey 2016 [3], it was found that 89.3 percent of internet users in Malaysia used a smartphone to access the internet. In addition, the percentage of smartphone ownership rose to 90.7 percent in 2015 compared to 74.3 percent in 2014.

Table 1. Growth of Internet Subscription and Smartphone Users

	2015	2016	2017
		(million)	
Fixed Broadband	2.8	2.5	2.6
Mobile Cellular	27.8	28.5	35.3
		(% of Penetration Rate)	
Broadband	99.7	99.8	117.3
Mobile Cellular	143.8	139.9	131.1
Smartphone User	87 .3	87.9	89.4

To increase cellular network capacity by deploying more base stations requires expensive capital expenditure (CAPEX) and operation expenditure (OPEX) [4]. Therefore, by leveraging on the existing radio access technology (RAT) is an interesting alternative and offers a promising solution. Data traffic from cellular networks could be offloaded to other existing RAT such as femtocells or WiFi and are referred to as mobile data offloading. Due to the abundance of WiFi infrastructure that is widely deployed by internet service providers (ISP) and residents, Wifi offloading is a viable solution to address the need of additional network capacity and at the same time providing wider network coverage possibly via WiFi features such roaming.

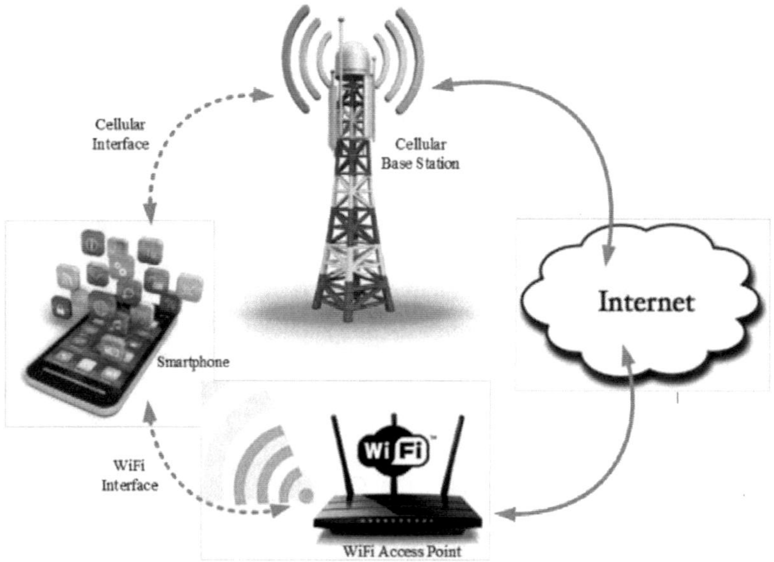

Figure 1. Mobile Data Offloading

Furthermore, it was announced by Telekom Malaysia (TM) which is the leading ISP in Malaysia that it has deployed over 10000 WiFi hotspots and is planning to increase its coverage [5]. Therefore, partnership between ISP and cellular network provider might be beneficial to both sides. Such implementation and collaboration need further understanding, feasibility studies and proper planning on consumer demand, as well as user's WiFi utilization pattern.

However, there is a lack of knowledge and information on the mentioned area, especially the one focusing on smartphone user's WiFi utilization. Hence, this paper aims to investigate WiFi utilization pattern in terms of temporal coverage, specifically in the context of Malaysian mobile data subscriber. [1] conducted a quantitative study on the performance of WiFi offloading by an experimental analysis of metropolitan area in South Korea. This research intends to simulate the study by focusing on the perspective of Malaysia's mobile user and understanding the influences of socio-economics group on WiFi offloading.

An Android based mobile application has been designed and developed as a tool to collect real data trace of user's daily-life pattern. This research extends [1] works by comprehensively investigating the socio-economics group influences on WiFi offloading. On WiFi utilization pattern, the study investigates the temporal coverage or the time portion users would like to spend in WiFi, as well as frequency of connection and properties between different groups.

All observations take into consideration the data behaviour during active hour and all-day connection. A comprehensive analysis of real data trace of smartphone's WiFi usage for18 days study, from 100 Android smartphone users has been studied. The main goal of the study is to observe the WiFi utilization details of smartphone users, as well as answering questions such as how much is the frequency and time portion in a day, that a smartphone user stays in WiFi coverage area? What time of the day that user use WiFi the most?

The measurement study is designed and conducted specifically for Android smartphone users which also has been categorized according to different socio-economics group. An Android based application called WiFi Analyzer has been designed and developed to collect real data traces of user's daily WiFi utilization. The research also considers the correlation of different socio-economics group on the several key parameters which is a method that has never been studied before.

WiFi offloading has been highlighted as a promising alternative that could assist cellular network provider in accommodating the ever-increasing demand of network capacity and network coverage. Therefore, it is necessary to further understand the current scenario in the consumer's market before any proper planning and deployment can be implemented. In this context, this work presents statistical observation, where outcome from the measurement study might be useful in understanding the bigger pictures of Malaysian internet scenario.

Additionally, the end users or consumers also might be interested in WiFi offloading due to economics reason, e.g., potential reduction of subscription fees or some value-added service with the same fees. Although there are some limitations in this study, it is hoped that the results of this study will be significant for future research endeavours. Findings of the study reveals new knowledge and information on the specified research area, which may provide insights and beneficial for cellular network provider, internet service provider and even regulatory body for the purpose of telecommunication network planning and development in Malaysia.

2. Related works

[1] claimed to perform the first trace-driven simulation using the measurement study in South Korea. The study recruited 97 volunteers from iPhone user community in South Korea and asked them to install DTap in their phones for the period of the measurement study. Findings reported the average of temporal coverage for all users are 70% for all day and 63% for active hour only.

From the other side, study by [6] reported only 11% of temporal coverage. Their work has a different approach where measurements were done only when a user is on a vehicle. Typically, users spend most of the time in the office or at home.

To the best of our knowledge, this is the first measurement study that carried out a real data trace of user's daily WiFi utilization pattern in the context of Malaysian cellular network subscriber. Using an application that has been designed and developed for Android platform to autonomously collect and recorded data, the study is specifically targeting on Android mobile devices such as smartphone and tab users.

Throughput, jitter, latency and QoS are the most common key parameters investigated in improving network performance, whether in fixed broadband environment or mobile networks infrastructure. It is found from the continuous literature search that most of the research carried out in Malaysia focus on mobile data networks and fixed broadband, and less research from wireless LAN, especially from the user's perspective. However, due to the high broadband demand, as well as to support government policy of NKEA, wireless LAN is an approach that should be given more serious consideration.

To provide internet services at Malaysia rural areas, [7] investigates application of a two-tier wireless network system with integrating wireless access and wireless backhaul as the evolution path of providing rural areas with Internet services. The first tier of the proposed network system is the wireless access network which provides interface to user accessing to Internet and enjoy services such as voice over IP, web browsing, and social network. The second tier is the backhaul network in which comprises of a number of wireless routers that carry traffic from various access networks to a gateway.

[8] conducted study on broadband internet performance in Malaysia, where the performance management is defined by the Quality of Services (QoS). The assessment was carried out among 5 regions in Malaysia with goal to identify the benchmarking in network performance. However, the study

was conducted for fixed line broadband where the assessment mostly was carried out using desktop and laptop users.

[9] conducted experiment to prove that single radio access unit (RAN) not only helps reducing MNO investment but it is also to improve network for coverage and quality compared to conventional radio deployment. The performance of the single RAN was studied in terms of coverage and quality compared to conventional radio. The specific tools which is called Nemo Outdoor is used to measure the coverage, quality and throughput via drive test. Important parameter and data captured are being process using Nemo Analyse and MapInfo Professional.

On the other side, [10] conducted research on the drivers of broadband in Malaysia and focusing on the identification of two relevant research stream covering broadband, which are adoption and continuance. Findings from the paper revealed that ISP, regulatory bodies, cellular network provider and application developer need to co-ordinate to support the adoption growth of broadband internet

3. Experimental setup

This section briefly explained the three phases which *Data Source, Data Understanding and Data Preparation* where each phase plays crucial roles in ensuring systematic data collection as well as reliability and validity.

A. Data Source

- *Location, Population and Android-based Platform*

To further study the WiFi temporal coverage, users from various levels are chosen in order to produce a variation of the sample. The participants are from various areas around Melaka, which has a diverse occupational background such as lecturers, private sectors worker, housewife, teachers, and students.

The accessible population that was selected consisted of public users and undergraduates from University Teknikal Malaysia Melaka with accessibility to WiFi through Android-based mobile devices. The rationale for choosing undergraduates is because they were in the age range of 20 to 24 years and, according to statistical data from the MCMC, the highest internet users are in the age group of 20 to 24 years old.

On the other side, a report by Device Atlas Mobile Report for Q2 2017 [11] reveals that Android as the leading mobile OS in Q1 2017. Figure 2 shows that Android OS dominated Malaysia mobile market share by 83.17% compared to only 14.33% by iOS.

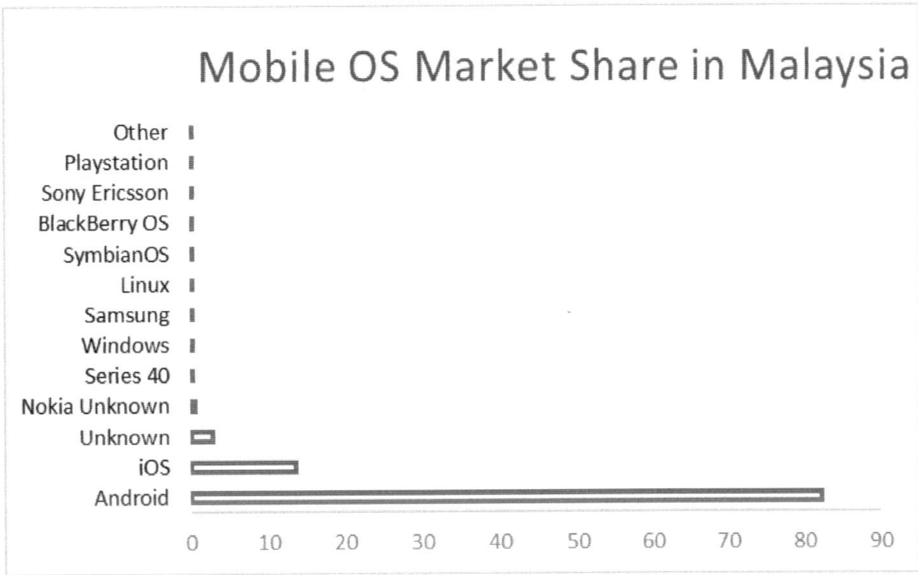

Figure 2. Mobile OS Market Share in Malaysia

- *Sample Size*

For the purpose of this study, Raosoft® software and Cochran's formula were utilised for the research. Raosoft® software is an efficient and user-friendly software that can easily be installed in any technological device including smartphones or tablets.

To determine the sample size using Raosoft® software, 10% of the margin error and a 95% confidence level were input parameters that were included. 100 participants from Android user community has been recruited and asked to install WiFi Analyzer on their phones. As the aims of the research are to investigate temporal coverage according to a different socio-economics group, participants have been categorized into two groups.

Group 1: The sample represents higher institutional students from University Teknikal Malaysia Melaka (UTeM). 25 students are from Faculty of Pengurusan Perniagaan (FPTT) located at Campus Jalan Hang Tuah, while another 25 students are from Faculty of Electronics Engineering and Computer Engineering located at Main Campus Durian Tunggal.

Group 2: The second group of 50 participants represents public users which comes from a diverse occupational background such as government and private sector employees, housewife, and teachers from various locations in Melaka.

- *Instrumentation*

The process of data gathering is done using autonomous Mobile Data Collection (MDC) method. MDC is a method of evaluating quantitative data using mobile phones, tablets or PDAs for programming or data collection. an Android smartphone platform called WiFi Analyzer has been developed. WiFi Analyzer tracks the data traces of WiFi activity in the background, saves the record in a text file and systematically upload the file to an FTP server.

The application captures all WiFi SSID available within its coverage area. The timer starts once the WiFi connection is established, and a log file is created in user's local storage. Otherwise, the application will keep checking the WiFi state. When the timer reaches three minutes, the application saves the current date and time stamp. The application also scans location and records the latitude and longitude of the current session of WiFi AP. Note that location only can be obtained if user enables location service on their mobile device.

The Android application has been developed using Android Studio tool with minimum Android SDK 21 (equivalent to Android 5.0 Lollipop) version, or newer is required. Data were sent to the server via file transfer protocol (FTP). Thus, Amazon web server (AWS) has been configured to receive data from users.

- *Procedure of Data Collection*

The application was ready to be distributed for the actual study after the validity and reliability were conducted and confirmed. In order to conduct the research in the respective faculties, the chronology of events leading to data collection was prepared. The description is as illustrated in Figure 3.

B. Data Understanding

Fraction of time that a user stays connected to a WiFi coverage area could determine offloading performance and is defined as temporal coverage. The key parameters that are crucial to measure temporal coverage are; connection start and end time, date of connection, daily total duration of WiFi connectivity for all day and active hour connection and daily tabulation of users with WiFi access at hourly period.

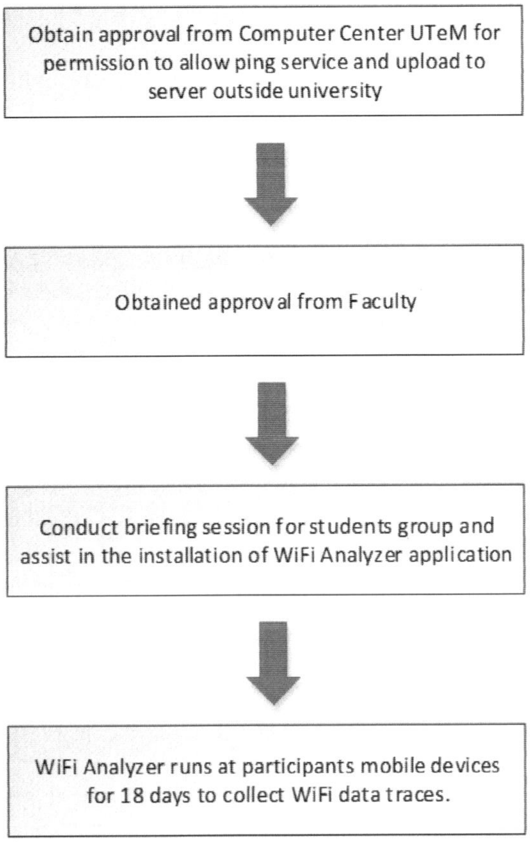

Figure 3. Chronology of Data Collection Process

C. Data Preparation

- *Data Cleaning and Extraction*

It is beneficial to understand the accuracy of collected data as well as correcting any errors or outliers in the data. Outliers are a single or small number of data values that are not similar to the rest of data set. The raw log file contained a massive data collection accumulated during the duration of the measurement study. There are a number of information that are not related to the scope of this research.

Therefore, a Python script has been developed to automatically extract the related information only. The Python script reads the raw log file line by line and only selects a predetermined field and organized the data into a table systematically. The script processes the wifi-log.txt file and produces an output file in excel .csv format. All necessary information can be extracted after identify the character surround the information. For example, referring to the string as shown in Figure 4 from the "wifi-log.txt", time is between string "time:" and "|(". Identifying these characters is actually the main method implemented in this script. All individual parameter has its own array to save the parameter in sequence order and tabulated into excel table.

```
Connection start time:18-09-2017-06-38-28|(102.2793452, 2.2809209)|
```

Figure 4. Identifying Characters in Python Script

4. Implementation of analysis

In analysis process, summary of tables and graphs can be used to convey information about the data. In this research, a summary table is used, which is a common way of understanding data. Summary tables will often show a count of the number of observation (or percentage) that have that particular value (or range).

Due to the comprehensive data collection, the data log for each participant can reach up to 10MB and obviously is a massive amount of data. The raw data files need to be sorted and organized before analysis can be carried out. Figure 5 shows the steps involved in data analysis procedure.

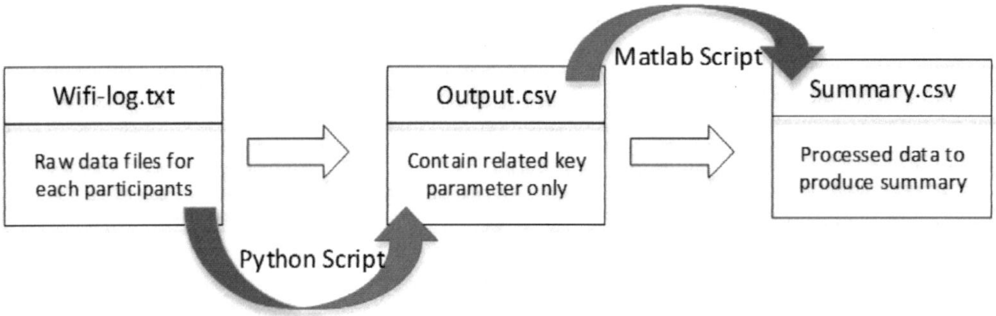

Figure 5. Data Analysis Procedure

The Python script is responsible to extract all the pre-defined information related to the research scope. However, the information still needs to be processed in order to produce a result and analysis. A MATLAB algorithm has been developed to do autonomous calculation and to analyse the temporal coverage. The program will first organize all the data according to the session of the connection, whether it is during active hour or all-day connection. Then, the algorithm calculates the duration of connection time for each day of the measurement study, which is used to measure the temporal coverage.

Temporal coverage is defined as time portion that user stays in a WiFi coverage areas and is one of the important key performance parameters in determining the performance of offloading. From the data collected during the measurement study, daily average temporal coverage could be determined for each participant.

The difference of time between all day reading and active hour (9.00 – 24.00) has also been highlighted. Temporal coverage can be determined by calculating the duration of connection time, where for each participant is the average percentage of connection duration for each day. Then, the percentage for the whole period of measurement study were averaged.

5. Results and discussion

A. Temporal Coverage

Figure 6 and Figure 7 shows the average of daily temporal coverage recorded by Group 1 and Group 2 participants for the duration of the measurement study. It also shows the plot of coverage recorded during the active hours (0900 to 2400) session. The red and blue bar represents the average of daily temporal coverage for each participant in the measurement study for all day and active hours respectively. Users are numbered as shown in X-axis and labelled as "User Index". The horizontal yellow and green line reveals the average reading for both sessions.

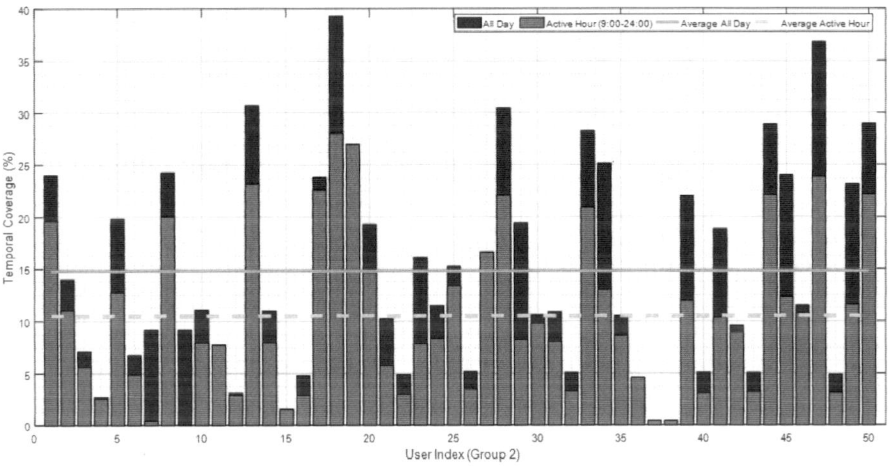

Figure 6. Temporal Coverage for Group 1 (Student)

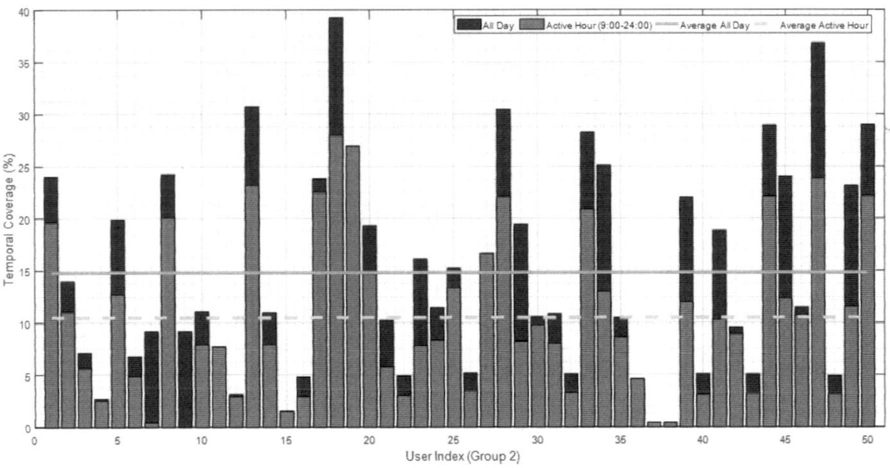

Figure 7. Temporal Coverage for Group 2 (Public)

As can be seen from the graph, the recorded trace of students in group 1 showed a higher temporal coverage for both categories all day and active hour connection. The average WiFi utilization for all day connection are 17.3% and 14.81% for group 1 and group 2 respectively, with only less than 3% difference. According to the demographics and socio-economics of internet user report by [12], more than 50% of Internet users in Malaysia are from the age group of 20-34 years old, which supported the finding that group 1 generally incline to access longer internet time.

Table 2 also shows that the recorded data trace has a quite significant standard deviation in the data sets from the mean value. Group 1 showed higher standard deviation of 15.6% during all day connection. The data indicates variety of user data pattern with minimum temporal coverage of only 1.7% to the maximum reading of 56.9%. However, the all-day median for both groups are almost similar with only 0.5% difference.

Table 2. Growth of Internet Subscription and Smartphone Users

	All Day (%)		Active Hour (%)	
	Group 1	Group 2	Group 1	Group 2
Min	1.731	0.4178	0.7	0
Max	56.9	39.29	31.12	28
Mean	17.30	14.81	10.8	10.49
Median	11.77	11.27	6.727	8.48
Std	15.6	10.11	9.572	7.812

For active hour, the data showed similar pattern of higher usage by group 1 compared to group 2 but with only slight difference of 0.31% for mean value. However, the median for group 1 and group 2 are 6.73% and 8.48% respectively.

Note that during active hour, the median for group 2 is higher than group 1. The data suggests that during active hour, public group spends more time using WiFi due to the availability and more reliable WiFi connection in the office or home. On the other side, the pattern showed that students usually spend more time in class or lecture during active hour and therefore has lower temporal coverage. [12] reported in Internet User Survey that place of study is the least popular place to access internet. The reason for lower temporal coverage for group 2 during active hour, also might be caused by the lack of coverage or stability of WiFi in campus during time spent on campus area.

B. User's Frequency Distribution

The frequency distribution presents information according to the number of observations reported for each ranges of values, for a particular variable. The observation for a range of percentage for both groups during all day temporal coverage is counted and plotted in the following histogram. User's frequency is plotted as density in Y-axis. The shape of the plot reveals trends, that is the user density for each ranges of temporal coverage percentage.

Figure 8 shows the histogram of all user (group 1 and group 2) temporal coverage during all day connection. The histogram indicates the number of users for each predetermined ranges of temporal coverage percentage. The result reveals that almost 50% from total participants recorded 10% of daily temporal coverage during all day connection.

IOP Conf. Series: Earth and Environmental Science **228** (2019) 012007 doi:10.1088/1755-1315/228/1/012007

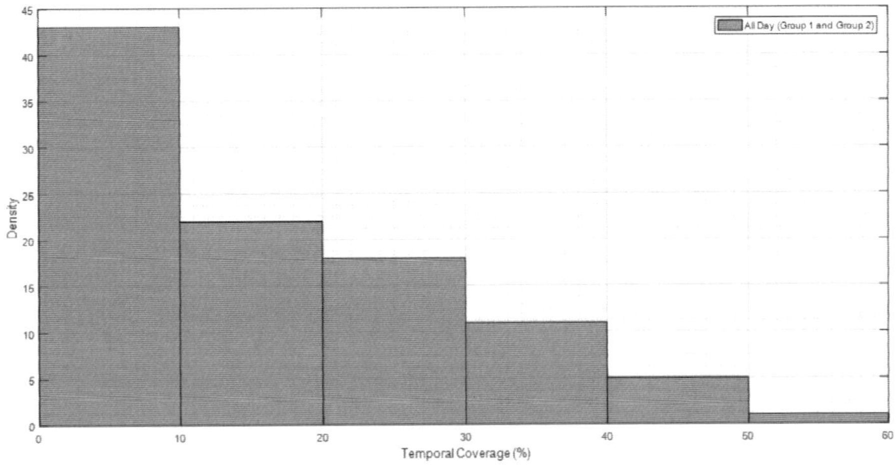

Figure 8. Combined Histogram for Group 1 and Group 2 (All Day)

C. Tabulation of User (Hourly)

Figure 9 and Figure 10 illustrates the average percentage of users with WiFi by hourly period for group 1 and group 2 respectively. It shows average percentage of users for each hour of the day, for the whole duration of measurement study.

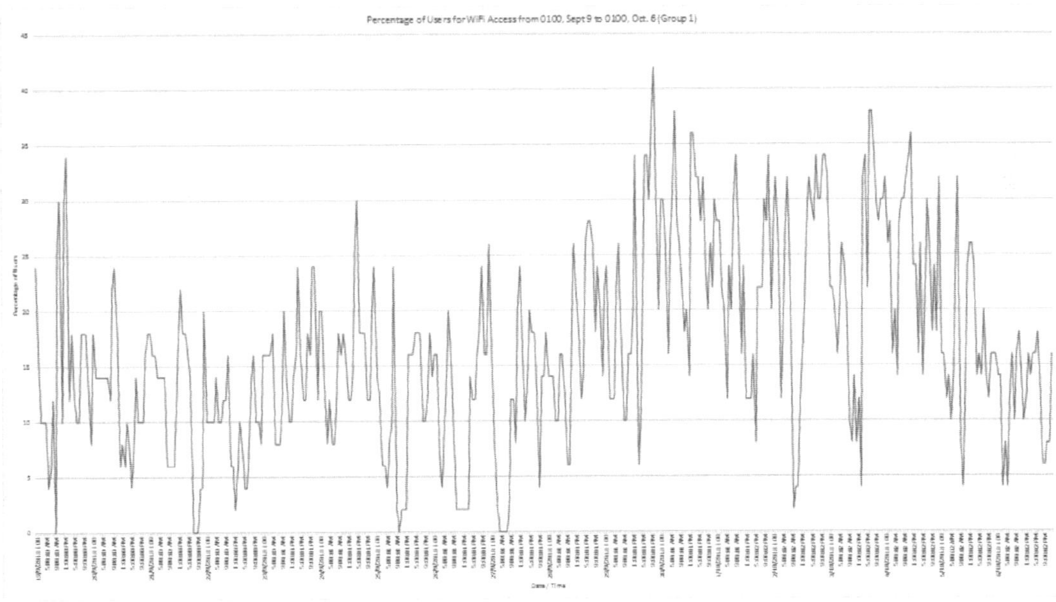

Figure 9. Percentage of users with WiFi access from 0100, Sept.18 to 2400 Oct.6 (Group 1)

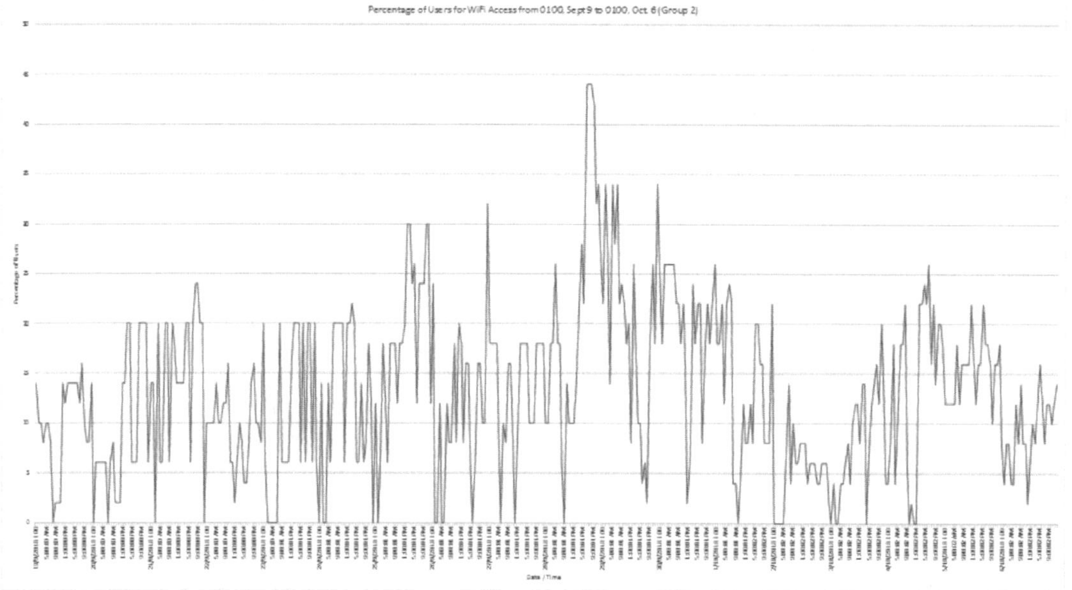

Figure 10. Percentage of users with WiFi access from 0100, Sept.18 to 2400 Oct.6 (Group 2)

The finding reveals that at any given hour of the day, the average percentage of users with WiFi access for group 1 is 21%. Percentage of group 1 user access is higher between 1300 to 0100 and the lowest access is between 0300 to 1200. This is most likely because students usually have class in the morning session and tends to access internet afterwards, whether at home, hostel or at campus. In addition, (Malaysian Communications and Multimedia Commission, 2017) reveals that place of study is the least favorite place to access WiFi.

On the other side, in average only 14% of group 2 users access WiFi by each hour of the day period. The finding reported that the highest WiFi usage for group 2 usually is between 1400 to 2300. Group 2 generally spends less time in accessing WiFi and the result is concurrent with the temporal coverage result as per highlighted in Table 2.

6. Conclusion

This paper presents a real trace of quantitative measurement study on the smartphone user's WiFi utilization in terms of WiFi temporal coverage in Malaysia. Two comprehensive datasets representing different socio-economics group were collected to further understand its influences in WiFi utilization. An autonomous method of data collection has been conducted using Android application, known as WiFi Analyzer which has been developed specifically for this study.

The finding reveals that the average temporal coverage for all participants are 16% for all day. It means the average daily WiFi usage for each user is only 4 hours. Group 1 spends more time using WiFi compared to Group 2, while both groups shows higher temporal coverage during active hour. The result is consistent with correlation analysis that shows group 1 has higher temporal coverage. The significant correlation is highly related to the difference between all day and active hour. In other words, higher difference between all day and active hour means higher correlation value and resulted in higher temporal coverage. In addition, histogram of user frequency distribution discovered that 50% from all users in both groups has average of only 10% daily temporal coverage.

Finding on tabulation of users by hourly-period indicates that average of 21% of group 1 access the internet at any given hour of the day, but the highest is between 1300 to 0100. Group 2 only has average of 14% user access at any hour of the day with high WiFi utilization between 1400 to 2300.

The finding indicates low WiFi temporal coverage for both groups of smartphone users. The result is contradicted with [12] data that reported 89.4% of internet users in Malaysia use smartphone as a preferred medium to access internet. But our finding reveals that smartphone users only use average of 4 hours daily, to access internet through WiFi networks. The possible explanation for the low WiFi temporal coverage is, users might access the internet via data networks provided by the cellular network provider instead of using WiFi networks. The theory is supported with MCMC data that reported increase in mobile data subscription to 35 million in 2017. If the situation continues, it will be difficult for cellular network provider to keep up with the high demand of network capacity.

Acknowledgments
The authors would like to acknowledge the support provided by Fakulti Kejuruteraan Elektronik dan Kejuruteraan Komputer (FKEKK), Universiti Teknikal Malaysia Melaka (UTeM). This project is funded by Short Term Grant (Projek Jangka Pendek PJP) in FKEKK, UTeM (PJP/2018/FKEKK/(3B)/S01615). This project is partially funded by Universiti Teknologi Malaysia (UTM) under Vote No. 04G19 and Ministry of Education (MOE) under 4F966 research grants (FRGS).

References

[1] Lee, K. et al. (2010) 'Mobile Data Offl oading : How Much Can WiFi Deliver ?', ACM CoNEXT, 21(2), pp. 536–550. doi: 10.1145/1921168.1921203.
[2] MCMC (2017b) Pocket Book of Statistics.
[3] MCMC (2017a) 'Network Performance Report 2016 Measuring broadband and voice performance', (January).
[4] Suh, D., Ko, H. and Pack, S. (2016) 'Efficiency Analysis of WiFi Offloading Techniques', IEEE Transactions on Vehicular Technology, 65(5), pp. 3813–3817. doi: 10.1109/TVT.2015.2437325.
[5] Telekom Malaysia (2018) Unifi Wifi. Available at: https://www.unifi.com.my/unifi-en/ijoin/product-details.page?productId=wifi (Accessed: 7 June 2018).
[6] Balasubramanian, A., Mahajan, R. and Venkataramani, A. (2010) 'Augmenting mobile 3G using WiFi', ACM International Conference on Mobile Systems, Applications, and Services (MobiSys), pp. 209–222. doi: 10.1145/1814433.1814456.
[7] Kwong, K. H. et al. (2011) 'WiFied Up Malaysia Villages : A Case Study of Using WiFi Technology to Increase Internet Penetration Rate in Malaysia Rural Areas', (October), pp. 7–12.
[8] Idris, M. F. B. M. et al. (2014) 'Broadband Internet performance (QoS measurement) view from home access gateway in Malaysia', Proceedings - 2014 5th IEEE Control and System Graduate Research Colloquium, ICSGRC 2014, pp. 147–152. doi: 10.1109/ICSGRC.2014.6908712.
[9] Ya, N. et al. (2014) 'Coverage and Quality Radio Analysis for Single Radio Access Network Technology in GSM', pp. 158–163.
[10] Lurudusamy, S. N. and Thurasamy, R. (2016) 'The Drivers of Broadband Internet in Malaysia', Proceedings - 6th International Conference on Computer and Communication Engineering: Innovative Technologies to Serve Humanity, ICCCE 2016, pp. 225–230. doi: 10.1109/ICCCE.2016.57.

[11] DeviceAtlas (2017) 'DeviceAtlas Web Intelligence Report Q2 2017', (August).
[12] Malaysian Communications and Multimedia Commission (2017) 'Internet Users Survey 2017 Statistical Brief Number Twenty-One', pp. 1–51. Available at: http://www.mcmc.gov.my

IOP Conf. Series: Earth and Environmental Science **228** (2019) 012008 doi:10.1088/1755-1315/228/1/012008

An Automatic Power Cut-off Appliance with Smart Battery Optimizer

Saipunidzam Mahamad, Muhammad Farid Mohd Zaid

Computer and Information Sciences Department, Universiti Teknologi PETRONAS, 32610 Seri Iskandar, Perak, Malaysia

saipunidzam_mahamad@utp.edu.my

Abstract. Appearance of new technology every single day in today's world is not some strange phenomena. The current trend shows new devices created to help and ease people's life, however, is still lacking in some factors which include Computing. Many had faced with issues of the battery lifetime, power drain due to overcharging and others. Even though there are some applications that claimed can help users to avoid the decreased of the problem but still users need to manually do the prevention such as unplugging the charger when the battery is full. Thus, this paper will present a development of an internet of things (IoT) device which can communicate with a mobile application to help people in handling their electronic appliances. The appliance integrated with an application that able to automate process, monitor the charging activities and apply mechanism to prevent an overcharging with some pre-setting value. Initial study received positive feedback thus support the development of this project. It gave beneficial to the society and simultaneously reduce the rate of producing the Li-ion battery waste in the world.

1. Introduction

Everyday people are inventing a new thing either as innovative product or enhancing technology to ease people's life from time to time. Invention on battery is one of phenomenon for all electronic appliance. In order for an electronic device to work or switch on, it will need a power supply such as a battery. In this new age, there are many types of battery that can be used to power up an electronic device and selection of the battery type also very crucial process for creating a new electronic device especially on how long the electronic device can be used for a day. Many of the inventors created a new electronic invention that will be using a battery that can be recharge when it has worn up as this method can indirectly attract the users to buy and use the device as they do not have to buy a new battery every time the battery is dead.

Energy conservation indicates to decreasing energy through utilizing less of an energy benefit. Energy conservation varies from effective energy employ, which refers to utilizing less energy for a constant service. Conserving energy gives researchers more opportunity come up with solutions and alternatives. In reality, all energy produced and utilized has an impact on the environment. Indeed, even energy from very normal sources affects the environment. To sustain the energy, is to create an energy without placing them in risk of getting expired or drained and can be used repeatedly. Sustainable energy ought

Content from this work may be used under the terms of the Creative Commons Attribution 3.0 licence. Any further distribution of this work must maintain attribution to the author(s) and the title of the work, journal citation and DOI.
Published under licence by IOP Publishing Ltd

to be broadly encouraged as it does not make any damage the environment and is accessible generally free of expense.

Batteries are the most widely recognized type of daily hazardous waste. However, the environmental effect of batteries is not constrained to the waste stream. Natural effects happen in the production, dissemination and end-of-life periods of the battery life cycle. Researcher agreed that a lead battery are key to a cleaner, greener future [1]. It is a most environmentally sustainable battery innovation and a pioneer in sustainable power source stockpiling arrangements. The spotless, sustainable power sources expected to help make a sustainable society.

A rechargeable battery is very good for all electronics device, as it will save user's money to buy a new battery. This benefit will also come with a deficit as that rechargeable battery may not last very long if the users always overcharge the battery. This occurrence always occurs due to hectic lifestyle as always forget to unplug the battery charger. The decreased battery life will really give a very bad experience when it comes to a place that do not have any power supply plug as a damaged rechargeable battery will always need the current to flow in it in order for it to power up the electronic device.

Even though, users will have the option to replace their rechargeable battery, but it will be quite costly. Some of the rechargeable battery have been permanently embedded inside the device which require some technical skills to replace it and give difficulties to users. The damaged battery may cause by overheating when charging for a long time or too many running applications. An excess heat that has been produced by that situation will damage the battery over time.

Therefore, this paper presents a study that integrate the development of a power cut-off appliance that integrate with application to monitor the charging activities. It applies some mechanism to send signal in between application and devices that will help users to maintain their original rechargeable battery for a longer time. The output of this research also will be a better solution for current mobile applications thus prevent the overcharging phenomena.

2. Literature Review

Battery is a device that have electrochemical cells that will be connected to external connections to act as a power source for electronic devices. According to Scrosati et al. [1], battery is a portable device that is capable of giving the stored chemical energy as electrical energy to power up any electronic devices. The battery has a high conversion efficiency with zero gaseous emission so that will make battery one of the green source of energy if compared to fossil fuel. Hannan et al. [2], battery is a high energy density, long lifespan and high efficiency energy storage that way better than the renewable energy like solar energy and wind energy. This is due to the battery properties itself make it better as it is portable and rechargeable plus it also easy to use.

There are many type of battery has been produced and the usage of the battery in any electronic devices is not a new phenomenon to people nowadays if compared to people who live in 1800 when battery first invented. According to Guo et al. [3], battery nowadays is widely used in any portable electronic devices and any new energy inventions such as the electric car. Zou [4] added that battery is commonly used throughout the globe in any devices that need electrical energy to power it up. All these statements have shown that battery nowadays is one of the most crucial things that is needed by human in order to survive in this new world.

According to Shen [5], batteries will always possible on undergoing a process called accelerated degradation because of its harsh charging/discharging cycle and high peak power. This situation will lead to the decrease of battery life time. In order to eliminate this kind of situation, there is a need to do a battery optimization technique to make sure the battery can last long. Alshurafa et al. [6], a battery optimization technique has successfully enhanced the battery life time up to 300%. The test has proven that with a battery optimization technique, it enhances a battery life time and at the same time conserve the energy of replacing the battery frequently.

Not only conserving the energy, battery optimization also can help users from any severely injured from burns associated with any battery-powered devices. If we did not take care of our smartphone battery correctly, the spontaneous battery explosion may happen [7].

Current battery optimization strategy that have been applied in laptop and smartphones is only by using the application that can give users tips on conserving their battery. There are no hardware or software that can help users optimizing their battery except by doing a manual job like plug out their electronic devices when their battery is full. This will give users an extra task whenever wants to charge their electronic devices. Due to certain circumstance such forgot to plug out their battery, it will lead to the situation of battery overheating and that situation may cause a decrease a battery lifetime as per discussion in an article by Guo et al. [8].

3. Methodology

3.1. System Architecture

Figure 1 shows the system architecture of the Smart Battery Optimizer device and mobile application. The medium of communication between the controller (device) and the mobile application installed in users' smartphones is by using the Bluetooth communication system. The mobile application will send a digital signal to the device whenever any one of the condition set by the have been met. Whenever users want to charge their smartphones, they will need to set the value of temperature and battery level that they want. After setting all the values, inside their mobile application, users just need to pair their smartphone with the device and everything will be done automatically. Users just need to make sure that their smartphones' charger is connected to the device so that the mobile application can controller the in and out of the electric current to the smartphone. The charging process of the smartphones will continue as usual as it will only stop when two condition that has been set earlier has been met. The first condition is when the battery level of the smartphones has reached the limit and the second condition is when the temperature of the smartphones has reached the maximum temperature set by the users.

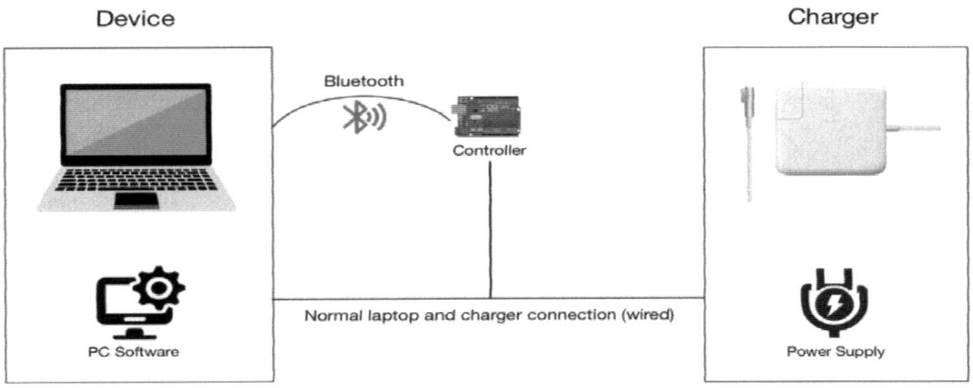

Figure 1. System Architecture for Smart Battery Optimizer

3.2. System Design

The are many battery optimizer applications that already exist in the market but based on my research, all of them need manual ways to preserve users' smartphones. As for now, there are no similar application in the market that have the same functionality as this application. The reason why this application is different from the others is due to the automatically stop the electric current from flowing from the power supply to the smartphones. Even the charging cable that users use will not have any current flowing inside when the controller does its job.

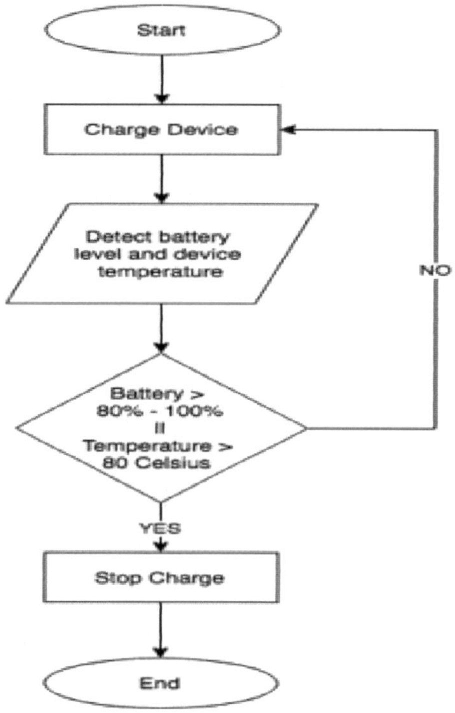

Figure 2. Smart Battery Optimizer Flowchart

Figure 2 shows the flowchart that explain more on how this device working. The process starts when the users want to charge their smartphones as the smartphones' battery has depleted. The users will need to set the limit of the smartphones' battery level and also the temperature as the controller will stop charging it receive a digital signal from the mobile application when the limit has been fulfilled. The mobile application installed in the smartphones will detect the battery level and also the temperature of the smartphones as it is running on the background. If the mobile application figures out that the smartphones has fulfil any one of the first and second condition, it will automatically stop the charging process. The charging process will start back if the software detects that the smartphones match either the third or fourth condition that is when the battery level of the smartphones has decreasing below the limit and also when the temperature of the smartphones decreasing below the maximum temperature set by the users. This process will loop until the users close the mobile application in their smartphones.

The usage of battery optimizer are needed to increase the battery lifetime, so user will have no problem to access their devices without any power source around them. This also will simultaneously reduce the rate of producing the Li-ion battery waste in the world as the Li-ion battery waste percentage always increase throughout the years.

4. Result and Discussion

A set of usability test has been conducted for a group of 10 respondents of users for the hardware device and software application. All the respondents come from diverse backgrounds such as students, professional workers and also full-time housewives. All the respondents have been given the hardware device with a smartphone that has been installed with the software application. After they test the product, all users have been given a link to complete an online survey that consist of six questions regarding their satisfaction and experience of using the product.

All the respondents have given positive feedbacks towards the hardware device and also the software application. Commonly, all users are very satisfied with the user interface of the Smart Battery Optimizer application as it is easy to use and easy to understand. The development of this product is to ease people's life in monitoring their smartphones during the charging process. It also can help users to extend their smartphone's life time from less than two years to four years or more.

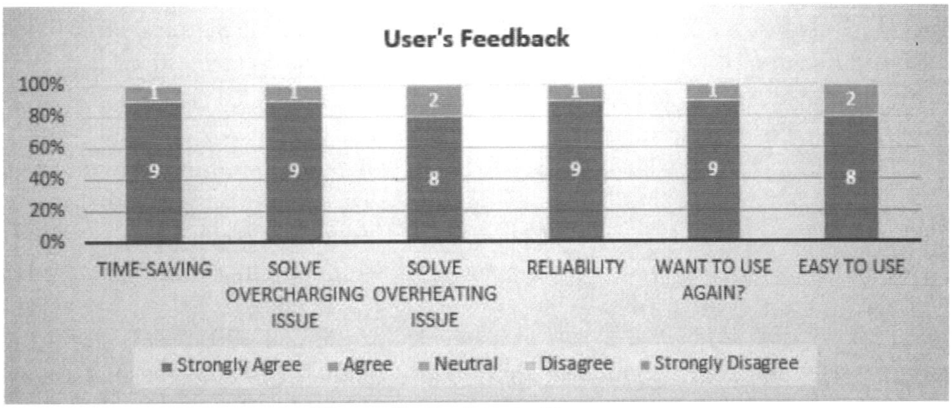

Figure 3. Participant Response

Figure 3 has illustrated that majority of the tester of the product agreed that this application and device can really save more time as it remove the time that we need to monitor whether our smartphones is fully charged or not. The 80%-90% of the testers agreed that this product can really solve the problem of overcharging and overheating of smartphones. 90% of the testers find that this application and device is really reliable, and they want to use it again after the test. All the testers have actually had the thought of straight away buy the product after the test as the product can really help to ease the daily routine. All the testers are very satisfied on the experience where they no need to monitor their smartphones when they charge it. They also find that this product also can really help them as it removes the feeling of worry as they charge their smartphones, as with this product, they will not face the overcharging and overheating situation.

5. Conclusion

Integrated Smart Battery Optimizer application both hardware and hardware is the solution to create a new and better way of controlling an electrical appliance. The development with a mechanism to automate the monitoring of charging activities would enable the process of cutting the power within the charging process. It will make electronic devices last longer and reduce waste of lithium ion battery as that will produced high impact pollution. By having this system, the battery lifetime will increase without any extra effort to plug out the battery every time it is fully charged. It improvise all the current applications in a way that all the manual job of getting users' smartphone battery optimize and the problems like overcharging situation could be solved.

IOP Conf. Series: Earth and Environmental Science **228** (2019) 012008 doi:10.1088/1755-1315/228/1/012008

6. References

[1] Scrosati, B., Hassoun, J., & Sun, Y.-K. (2011). Lithium-ion batteries. A look into the future. Energy & Environmental Science, 4(9), 3287.

[2] Hannan, M. A., Lipu, M. S. H., Hussain, A., & Mohamed, A. (2017). A review of lithium-ion battery state of charge estimation and management system in electric vehicle applications: Challenges and recommendations. Renewable and Sustainable Energy Reviews, 78, 834-854.

[3] Guo, Z., Qiu, X., Hou, G., Liaw, B. Y., & Zhang, C. (2014). State of health estimation for lithium ion batteries based on charging curves. Journal of Power Sources, 249, 457-462.

[4] Zou, Y., Hu, X., Ma, H., & Li, S. E. (2015). Combined State of Charge and State of Health estimation over lithium-ion battery cell cycle lifespan for electric vehicles. Journal of Power Sources, 273, 793-803.

[5] Shen, J., Dusmez, S., & Khaligh, A. (2014). Optimization of Sizing and Battery Cycle Life in Battery/Ultra capacitor Hybrid Energy Storage Systems for Electric Vehicle Applications. IEEE Transactions on Industrial Informatics, 10(4), 2112-2121.

[6] Alshurafa, N., Eastwood, J., Nyamathi, S., Xu, W. J. L., Jason, & Sarrafzadeh, M. (2014). Battery optimization in smartphones for remote health monitoring systems to enhance user adherence. ACM International Conference Proceeding Series. 2014, 1-4.

[7] Mankowski, P. J., Kanevsky, J., Bakirtzian, P., & Cugno, S. (2016). Cellular phone collateral damage: A review of burns associated with lithium battery powered mobile devices. Burns, 42(4), e61-64.

[8] Guo, L. S., Wang, Z. R., Wang, J. H., Luo, Q. K., & Liu, J. J. (2017). Effects of the environmental temperature and heat dissipation condition on the thermal runaway of lithium ion batteries during the charge-discharge process. Journal of Loss Prevention in the Process In-dustries, 49, 953-960.

IOP Conf. Series: Earth and Environmental Science **228** (2019) 012009 doi:10.1088/1755-1315/228/1/012009

Performance Analysis of Real Time Image Processing for Lightning Event Using Cython and Python Programming Languages

S. H. Mun[1], M. R. Ahmad[1], R. F. Malik[2], M. R. M. Esa[3], M. H. M. Sabri[1], D. Periannan[1], B. Y. Seah[1], S. A. Mohamad[2], V. Cooray[4], A. A. Alkahtani[5] and M. Z. A. AB. Kadir[5]

[1]Atmospheric and Lightning Research Lab, Centre for Telecommunication Research and Innovation (CeTRI), Fakulti Kejuruteraan Elektronik dan Kejuruteraan Komputer, Universiti Teknikal Malaysia Melaka, Hang Tuah Jaya, 76100 Durian Tunggal, Melaka, Malaysia.

[2]Faculty of Computer Science, Universitas Sriwijaya (UNSRI), Inderalaya, Sumatera Selatan, Indonesia

[3]Institute of High Voltage and High Current (IVAT), School of Electrical Engineering, Faculty of Engineering, Universiti Teknologi Malaysia (UTM), 81310 Skudai, Johor Bharu, Malaysia

[4]Ångström Laboratory, Division for Electricity, Department of Engineering Sciences, Uppsala University, Box 534, S-75121, Sweden

[5]Universiti Tenaga Nasional (UNITEN), Jalan IKRAM-UNITEN, 43000 Kajang, Selangor.

Email: riduan@utem.edu.my, rezafm@unsri.ac.id

Abstract. Due to the uncommon and unpredictable phenomenon, lightning event become one of the challenging research topics for the scientist. The main thing that most capturing their attention is the lightning detection method which can improve the public safety by pinpointing the location of the lightning event and sending out the warning notification. Currently, there are lots of researcher proceed with the study related with the relationship between the pattern of the lightning and the electromagnetic signal received. During the research process, they would like to capture a clear lightning image from the cloud to proceed with the study of characteristic of the pattern of the lightning and electromagnetic field. The most common method for the researcher to detect and capture the lightning image is using the high-speed camera with a flash sensor. But this method is extremely expensive and has its own limitation on mobility part. Therefore, this paper is to develop and evaluate the suitable algorithm and programming language onto Raspberry Pi platform and act as an upgraded choice to replace high speed camera with affordable cost, high mobility but less accuracy compares to high speed camera.

Content from this work may be used under the terms of the Creative Commons Attribution 3.0 licence. Any further distribution of this work must maintain attribution to the author(s) and the title of the work, journal citation and DOI.
Published under licence by IOP Publishing Ltd

1. Introduction

Lightning is an uncommon and unpredictable phenomenon that happened in real life. Due to the sudden electrical discharge in a very short duration, it occurred for a very quick and short time [1]. Because of that, it is so hard for the people to get a clean shoot for a complete and perfect strike of lightning and analysis how it grows [2].

Commonly, people will use Digital Single-lens Reflex (DSLR) camera which have better International Organization for Standardization (ISO) setting to capture the appearance of lightning that only occur in short duration [3]. However, most of the DSLR camera does not have any Central Processing Unit (CPU) to do real time processing. Besides, this method is extremely high cost for researcher.

Therefore, this paper is to study and evaluate the suitable algorithms of image processing for lightning detection. Besides, all the algorithms will be coded by using two programming languages which are Python and Cython programming language. All the final data for the frame per seconds and the processing delay between two programming language will be analysed. Besides, the frames per second based on the different programming libraries also will be analysed. The pattern of the lightning image will be compared to the electromagnetic signal that received by lightning sensor installed at our lab.

2. Methodology

A. Real Time Image Processing

The main reason to proceed this real time image processing is to detect and capture the lightning image that appear on the camera lens. Before the pre-image processing process, there are several things that need to be defined and cleared out such as the save file location, the detection method, internal camera setting and capturing speed. For this process, OpenCV library and PiCamera library is applied by using Python programming language [4].

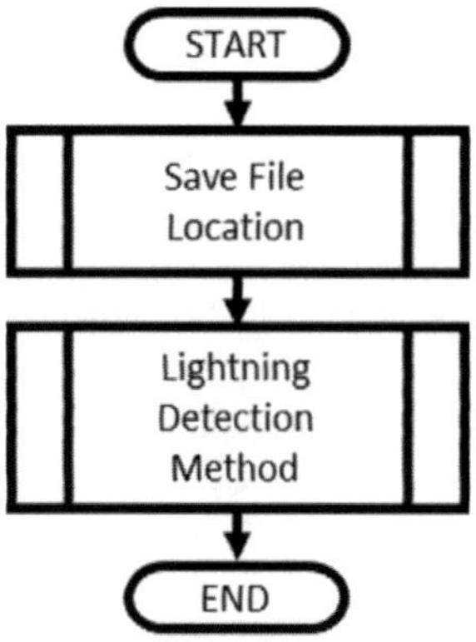

Figure 1. Flow chart for lightning detection system

Figure 1 shows the overall flow chart for the real time lightning detection system. This real time lightning detection system only focus on two main functions which are save file location function and lightning detection method function.

B. Save File Location

Before start capturing the lightning image that detected by the real time image processing system, save file location must be defined. Save file location is used to store the capturing results. Therefore, inside the system must create a new directory for the save file location automatically. In this process, the operation system itself must check whether the file had already existed in the directory. If the folder does not exist, operation system must create another new directory folder automatically. If the folder exists, it proceeds with next function.

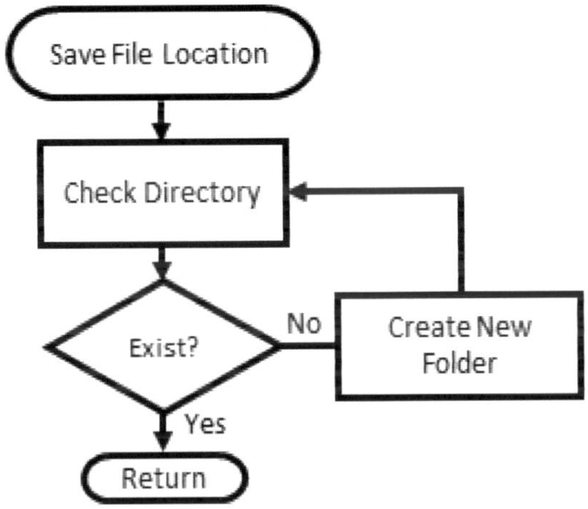

Figure 2. Flow chart for save file location function

C. Lightning Detection Method

For lightning detection method function, it used the pixel value to recognize lightning event. Once the pixel value exceeds more than the threshold value will be recognized as lightning event. After detecting the exceeding pixel value, this real time image processing system will start capturing image until the pixel value is less than the threshold value [5]. Those capturing image will be storing at the folder directory.

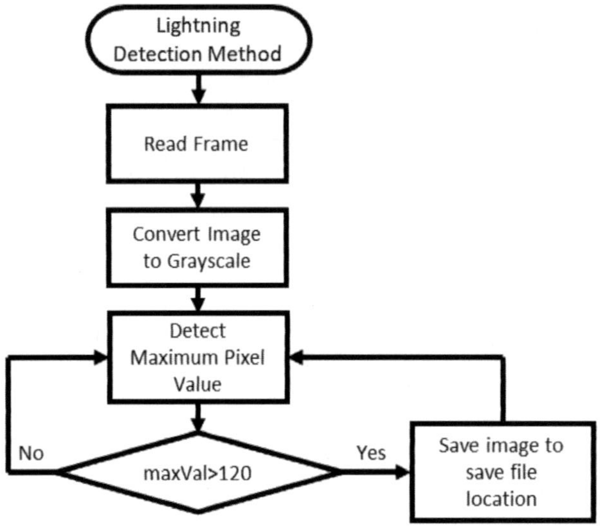

Figure 3. Flow chart for lightning detection method function

Figure 3 shows the flowchart for the lightning detection method for this real time image processing application. This lightning detection method include the conversion of grayscale image and detection for the pixel value [6].

3. Results and analysis

For this paper, the frames per second rate and processing delay are main objectives. Therefore, this paper is comparing the frames per second and processing delay with different perspective such as programming language and programming libraries. But because of the aim of this paper is focusing on lightning event, the structure and pattern of the lightning event also one of the consideration which will be used to compare with the electromagnetic field that capture through the PicoScope software.

A. Analysis of frames per second based on programming language

This paper analyzed the frames per second that based on the programming language that used for the real-time image processing system. There are two differences programming languages that used in this paper which are Python and Cython programming language. By using the same operating system and system flow for this real-time image processing system, compare the number of results based on the saving image files for one second. This testing will take the average number for 15 seconds.

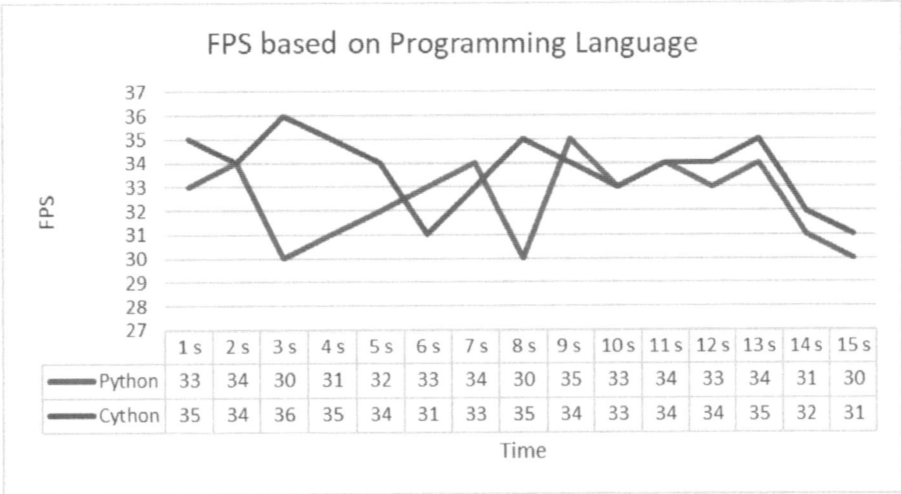

	1 s	2 s	3 s	4 s	5 s	6 s	7 s	8 s	9 s	10 s	11 s	12 s	13 s	14 s	15 s
Python	33	34	30	31	32	33	34	30	35	33	34	33	34	31	30
Cython	35	34	36	35	34	31	33	35	34	33	34	34	35	32	31

Figure 4. Graph of frames per seconds based on programming language

Based on the Figure 4 showed, the average number of the frames per second for the Python programming language is 32.47 frames. However, for the Cython programming language, the average number of frames per second is 33.73 frames. It is only slightly different between those two programming languages. This is because of the library that used in this paper is OpenCV library which is the library that very similar with C programming language. Therefore, we believe that OpenCV brings the benefits of faster processing speed when we used it as the system library.

B. Analysis of frames per second based on programming library using Python programming language

This paper also analyzes on the effect of the programming language on the results of the system based on the frames per second status. In this paper, two different libraries are used to analyze the different between their processing speeds. This real-time image processing system used Python as the programming language and apply each program library. By using the same operating system and system flow for this real-time image processing system, compare the number of results based on the saving image files for one second. This testing will take the average number for 15 seconds.

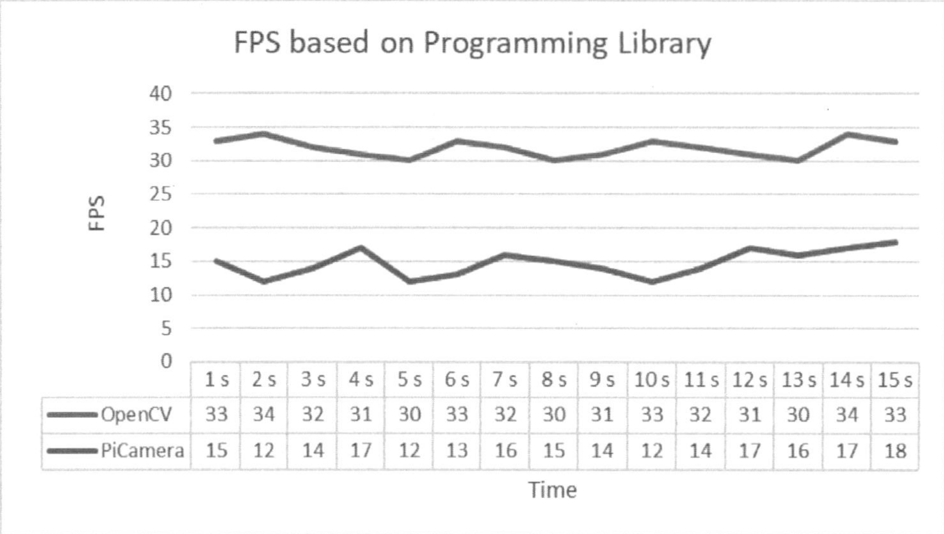

	1 s	2 s	3 s	4 s	5 s	6 s	7 s	8 s	9 s	10 s	11 s	12 s	13 s	14 s	15 s
OpenCV	33	34	32	31	30	33	32	30	31	33	32	31	30	34	33
PiCamera	15	12	14	17	12	13	16	15	14	12	14	17	16	17	18

Figure 5. Graph of the frames per seconds based on programming library

C. Analysis of the relationship of electromagnetic field with pattern of lightning event

Due to the lightning event itself is a high-current electrostatic discharge, every negative cloud to ground lightning strike will begin with a downward discharge which also known as leader stroke. When the leader gets closer to the ground, the tips of the leader will generate a strong electric field by connecting the upward discharge from the ground and then immediately followed the return stroke. Therefore, this paper captured the lightning event image with the real-time timestamp on it, the result of the lightning image can be identified which characteristic of the lightning by comparing with the pattern or structure of the lightning.

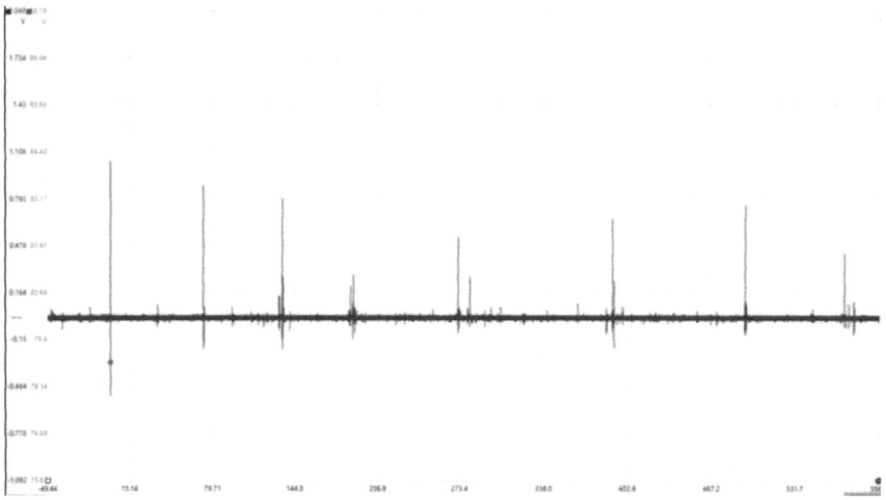

Figure 6. Overview of electric field that generated from Picoscope

Figure 6 shows that the overview for the electrical field that capture from Picoscope software. Based on the waveform from the graph, it can explain more details on the structure of the lightning event by knowing when the step leader and the return stroke is occurred. By further analysis the electrical field waveform from the Picoscope, this waveform can be used to synchronize with the lightning image that capture and do the analysis based on the exact time value. Figure 7 shows the first return stroke that zoom in from the overview of the electrical field that Picoscope captured.

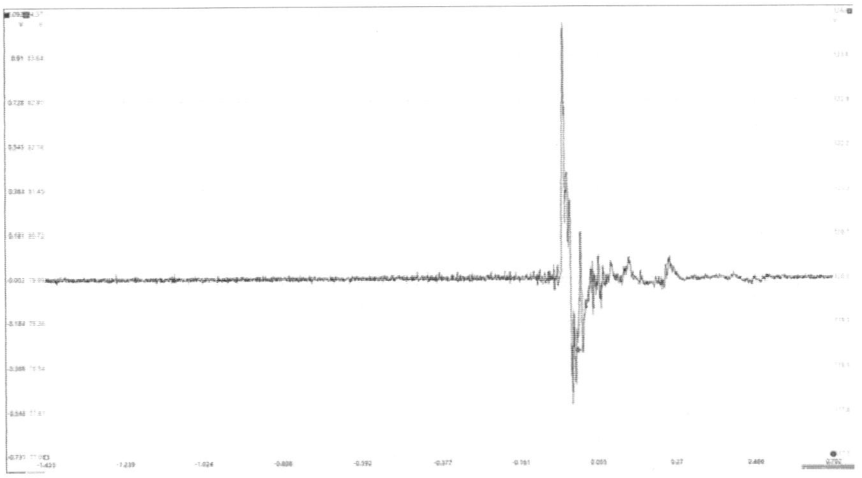

Figure 7. The first return stroke of the lightning event

Comparing the result of the lightning event from Picoscope, the electromagnetic waveform will be comparing with the actual lightning event image by synchronizing timing for both based on their timestamp.

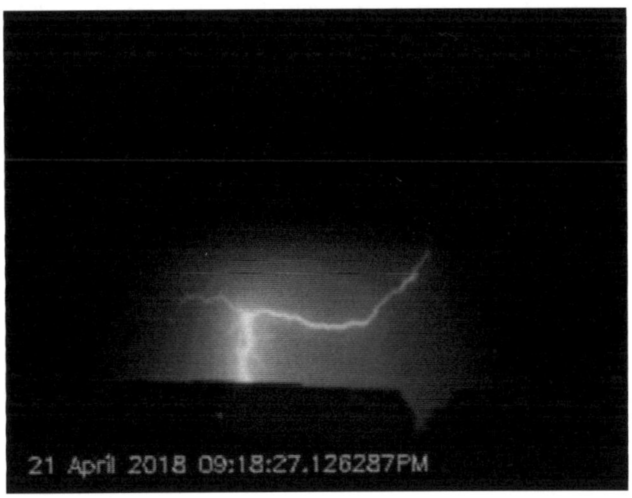

Figure 8. Image of first return stroke of the lightning event

By comparing with the timestamp of the lightning image that captured and the Picoscope results that we captured, we can clearly see that the lightning is in the brightest moment which also known as the return stroke moment. The branches of the lightning are started to branch out.

4. Discussion

This paper is an upgraded version to replace the DSLR camera with more affordable cost, high mobility and further image processing for the output capturing image. Therefore, this paper develops a real time image processing system which can detecting and capturing lightning event itself. Based on the results, Raspberry Pi 3 is a suitable minicomputer that can be use as the real time image processing system.

After analysis all the results, the most suitable programming that uses for this real time image processing system is Python programming language. This is because Python programming language is easier to build a real time image processing system compare to Cython programming language [7]. Based on the experimental results, even though Cython programming language is slightly faster than Python programming language, the limitation of research source for Cython programming language is one of the main reasons that Python programming language is easier to code.

For the programming library, OpenCV library is preferable compared to PiCamera Library. Mainly is because of the processing result for OpenCV library is almost two times faster than PiCamera library based on the experimental results. The reason that OpenCV library is processing faster than PiCamera library is because of OpenCV library is an optimizing C language library which is almost like C language coding [8]. Therefore, the processing time for OpenCV library is faster compare to PiCamera library.

5. Conclusion

Throughout all the research and troubleshoot part, this project has successfully build the real time lightning detection system with the suitable image processing method for analysis the lightning object. After analysis the experimental results, processing delay for the Cython programming language is lesser compare to Python programming language. But this statement must depend on which programming library that used. For OpenCV library, the processing time for both programming languages are almost the same due to OpenCV library is an optimizing C language library. In a conclusion, all the objectives are achieved.

Acknowledgments
The authors would like to acknowledge the support provided by Fakulti Kejuruteraan Elektronik dan Kejuruteraan Komputer (FKEKK), Universiti Teknikal Malaysia Melaka (UTeM). This project is funded by Short Term Grant (Projek Jangka Pendek PJP) in FKEKK, UTeM (PJP/2018/FKEKK/(3B)/S01615). This project is partially funded by Universiti Teknologi Malaysia (UTM) under Vote No. 04G19 and Ministry of Education (MOE) under 4F966 research grants (FRGS).

References

[1] A. C. V. Saraiva, M. M. F. Saba, O. Pinto, K. L. Cummins, E. P. Krider, and L. Z. S. Campos, "A comparative study of negative cloud-to-ground lightning characteristics in São Paulo (Brazil) and Arizona (United States) based on high-speed video observations," J. Geophys. Res. Atmos., vol. 115, no. 11, 2010.

[2] C. V. Boys, "Progressive lightning: A new stereoscope," Nature, vol. 131, no. 3310, pp. 492–494, 1933.

[3] R. C. Gonzalez, Digital Image Processing, vol. 14, no. 3. 2002.

[4] "Programming libraries," 2000. [Online]. Available: http://www.math.uni-leipzig.de/pool/tuts/FreePascal/prog/node13.html. [Accessed: 15-Apr-2018].

[5] J. Manasa, J. T. Pramod, S. A. K. Jilani, and S. J. Hussain, "Real Time Object Counting using Raspberry pi," vol. 4, no. 7, pp. 540–544, 2015.

[6] Bibek Joshi, "What is the difference between a black and white and grayscale image?," 2017. [Online]. Available: https://www.quora.com/What-is-the-difference-between-a-black-and-white-and-grayscale-image. [Accessed: 09-Apr-2018].

[7] "The Performance of Python, Cython and C on a Vector," 2014. [Online]. Available: http://notes-on-cython.readthedocs.io/en/latest/std_dev.html. [Accessed: 15-Apr-2018].

[8] "About OpenCV." [Online]. Available: https://opencv.org/about.html. [Accessed: 16-Apr-2018]

IOP Conf. Series: Earth and Environmental Science **228** (2019) 012011 doi:10.1088/1755-1315/228/1/012011

Utilization of big data analytics in socio-economic impact evaluation on climate response measures

W K Wong

Chinese University of Hong Kong, Hong Kong

maxwongwk@gmail.com

Abstract. For the past decade, the conversation about climate change has been gaining prominence in the social and economic discourse in Malaysia. This is evident when Malaysia responded positively to the Paris Agreement by committing to reduce Greenhouse Gas emissions by 45% by 2030 in relation to the country's 2005 GDP. The Paris Agreement is a multilateral environmental agreement under the United Nations Framework Convention on Climate Change (UNFCCC) that compel all signatory countries to focus on greenhouse gas emissions mitigation, adaptation and finance starting in year 2020 as part of the global response in addressing climate change threats. Currently, UNFCCC has acknowledged the need to establish a forum to address the impact of response measures implemented by countries that will be undergoing decarbonizing development in terms of economic diversification and just transition (creation of decent and quality job). This paper will explore the challenges and opportunities to utilise big data analytics in socio-economic models when assessing the impact of a country's response measures policy implementation.

1. Introduction

In 1997, the Kyoto Protocol was adopted by 192 Parties with the aim to reduce 5% Green House Gases (GHG) from the 1990 levels bound to the developed countries. Following that, a multilateral environmental agreement under the United Nations Framework Convention on Climate Change (UNFCCC) known as the Paris Agreement was adopted by the international community in 2015 with the aspiration to maintain the global temperature rise this century well below 2 degrees above pre-industrial levels. The agreement compels all signatory countries (called Parties) to focus on greenhouse gas emissions (GHG) mitigation, adaptation and finance begging in year 2020 as part of the global response in addressing climate change threats.

Malaysia responded positively to the Paris Agreement by committing to reduce Greenhouse Gas (GHG) emissions by 45% by 2030 in relation to the country's 2005 GDP. This target was set with 35% on unconditional basis and 10% on conditional basis upon receipt of climate finance funding, technology transfer and capacity building from developed countries. The main GHG emission contributors in Malaysia are the energy industries, transport, manufacturing industries and industrial processes, waste and the agriculture sector. In addition, Malaysia also committed to maintaining at least 50% level of forest and tree conservation.

Content from this work may be used under the terms of the Creative Commons Attribution 3.0 licence. Any further distribution of this work must maintain attribution to the author(s) and the title of the work, journal citation and DOI.
Published under licence by IOP Publishing Ltd

In order to achieve the ambitious goals as outlined under the Paris Agreement, the UNFCCC secretariat was setup to support all stakeholders involved in the climate change negotiation sessions. The Conference of the Parties (COP) is convened annually to allow Parties to participate in various plenary sessions, preparatory meetings and thematic consultations to negotiate on wide ranging issues with the purpose to advance the work programme for the implementation of the Paris Agreement. The on-going negotiations being discussed extensively were generally divided into several key areas which include mitigation, adaptation, finance, technology, capacity building, economic instruments, transparency of action and support. Generally, the developed countries, which are often the biggest carbon emission contributors, are expected to bear the heavier burden (providing fund, technology transfer and capacity building) in addressing climate change.

2. Discussion

2.1. The Scope of Response Measures on Climate Change

The Kyoto Protocol committed member countries to strive to minimize the possible adverse economic, social and environmental impacts on other countries, particularly the developing countries, when addressing climate change. The response measures were further addressed in the subsequent Bali Roadmap process, the Cancun Agreements and the Durban Outcome before finally institutionalized for the discussion in the Paris Agreement, which provide a basis for addressing the impact of the implementation of response measures. To date, the modalities, work programme and functions under the Paris Agreement of the forum on the impact of the implementation of response measures have placed a heavy emphasis on economic diversification/transformation and just transition of workforce and creation of decent and quality jobs.

The International Centre for Trade and Sustainable Development or ICTSD recommended that forum consider exploring the impact of response measures implementation namely 'green protectionism' that could create trade distortion due to emissions trading schemes, cross border carbon tax adjustments (BCA), national subsidies of green technologies and measures, carbon footprint and green labelling schemes and regulation of international transport [1]. Govindasamy shared similar concerns by highlighting that the climate response would affect trade and competition conditions across different sectors that may not be compatible with WTO rules, emphasising the need to develop a tool-box to assess the impact of response measures encompassing green jobs, emissions trade, carbon tax, energy efficiency standards, green labelling and subsidies [2].

To better evaluate and understand the consequence of response measures, this article will explore the data linkages and possible deployment of big data analytics between the following climate response measures with trade and industry developments:

2.1.1. Green Jobs Potential.

The International Labour Organization (ILO) has defined green jobs as occupations that serve to preserve, restore and enhance environmental integrity. The US Bureau of Labour further defined renewable energy, energy efficiency, pollution reduction and removal, GHG reduction, recycling, natural resources conservation and environmental compliance, education and training related goods and services as part of green labour force that are essential to a growing sustainable economy.

To date, the Input-Output (IO) analysis and Social Accounting Matrices (SAMs) are the most widely used empirical tools in assessing green jobs creation. These models would employ a matrix/table to list relevant sub-sectors by linking outputs from one sector as an input to another. By evaluating the labour intensity of an identified sector in a particular value chain, the model would be able to determine the impact of the increase in green service/ product on employment. Furthermore, the empirical data obtained from I-O analysis or SAMs could be carefully designed by combining with a series of economic equations to fully explore the effect of climate change response measures on future green employment scenario [3].

The data sets involved in a typical I-O analysis for job measurement would entail sales and purchase made by a specific sector (total purchase value per unit production), GHG emissions, fluctuating demand of green services/products, number of employees per sector and average wage rate [4]. However, Bowen and Kuralbayeva had cautioned that green policies may affect labour markets indirectly through supply chains and market demand changes such as elimination of non-green jobs from polluting industries and rising cost of employment for green compliance purpose [5].

2.1.2. Green Labelling

Green labelling or energy-efficiency labelling summarised a manufactured product's energy performance, thus allowing consumers to make an informed purchase while indirectly encouraging competition among manufacturers and retailers to innovate and market the most energy-efficient products. As an effort to encourage economic sustainability, some governments have considered imposing energy-efficiency standards by prohibiting the sale of products that are unable to fulfil the minimum standards threshold [6].

Cohen and Vandenbergh listed the following five main principles for an efficient green labelling screening methodology [7]:

a) Able to identify goods where changes in consumption (either substitution or reduced usage) would reduce carbon emissions significantly;
b) Able to account for the costs involved for information gathering;
c) Able to capture the carbon foot print of each step in the product's life cycle including production, transport, storage, sales, consumption and disposal.
d) Able to accommodate the behavioral plasticity of consumers purchasing a carbon label product with a new implicated cost; and
e) Able to account for those goods (e.g. polluting vehicles) that would greatly incentivise firms to pursue green labelling.

The illustration below highlighted the full extent of the measurement scope for green labelling specifically environmental analysis (GHG emissions), utility analysis (energy consumption) and manufacturing analysis (employment and cash flow).

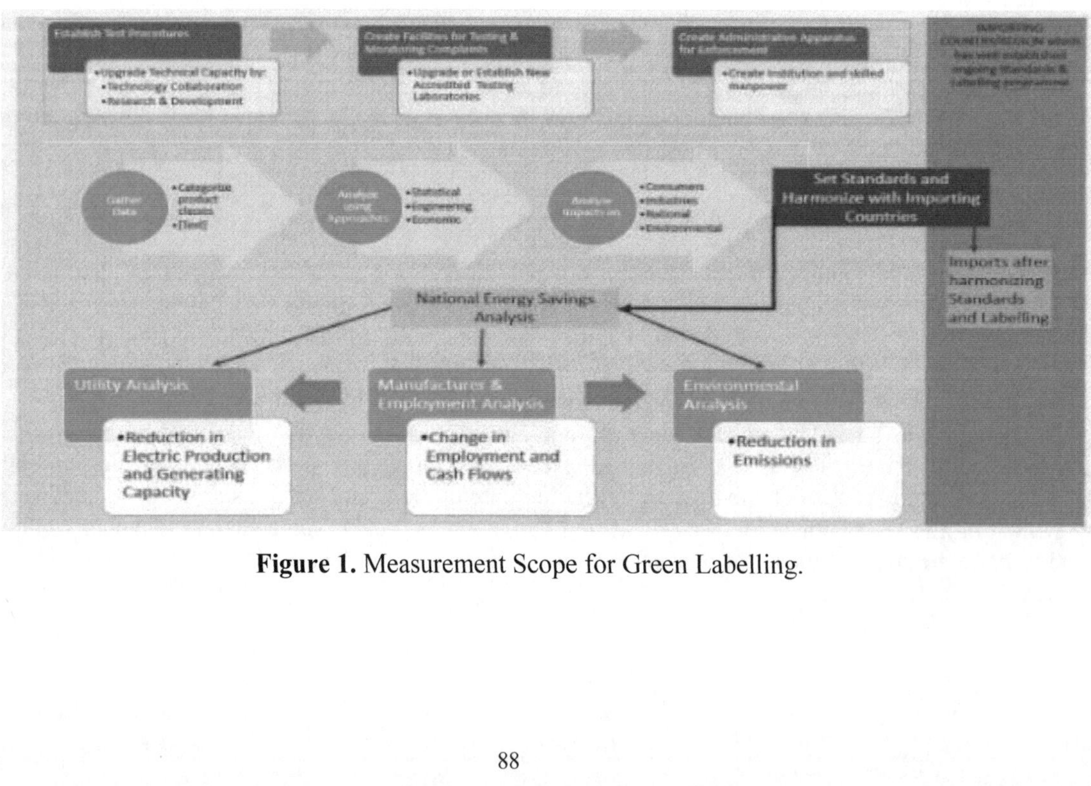

Figure 1. Measurement Scope for Green Labelling.

2.1.3. Cross Border Impact on Trade

According to IISD, there has been little progress or experience to date concerning the assessment of cross-border impact following the response measures implementation including carbon cap-and-trade schemes and green tariffs [8].

The United States Environmental Protection Agency defined cap-and-trade schemes or 'emissions trading' as one of the environmental protection approach by placing a limit (cap) on pollution for individual emissions sources and authorising tradable allowances that are equal to the emission limit of pollutant. The approach is market based in nature thus allowing the emission sources to freely buy and sell the allowance depending on their own compliance priority [9]. The assessment parameter for the scheme is mainly categorized into three areas namely:

a) Environmental Assessment – Evaluation of emission, ambient air concentrations, atmospheric pollutants related data.
b) Economic Assessment – Evaluation of environmental preservation, human health benefits and costs of compliance.
c) Market Assessment – Evaluation of allowance activity (e.g. number of transactions that occur) to ascertain economic benefits and costs of emission reduction target.

Green tariffs encourage customers to procure renewable energy in a collaborative arrangement between designated users and utility company, either through a subscription mechanism or direct purchase agreement for lower cost of electricity. Belcher and Sunkara elaborated that an effective green tariff framework would entail the following contracting and rate mechanism [10]:

a) Framework for efficient cooperation – Users will be involved more in the negotiations on pricing terms via sleeved power purchase agreements, providing financial certainty for the utility company to pursue renewable facilities financing project.
b) Phasing and renewable intermittency – Utility company will align the actual cost, with real time or seasonal variability, to reflect the intermittent nature of renewable energy sources.
c) Cost recovery – Users would be required to pay their fair share of the renewable resource cost with basic energy rate (based on consumed energy) and separate 'surge demand charge' to cover additional costs as well as increased rate adjusting to new technology adoption/deployment.

2.2. The Role of Big Data Analytics on Response Measures Implementation

Big data refers to large data volume, often unstructured, that inundates an organisation on daily basis. The big data concept first gained prominence in 2001 when Laney articulated and defined big data with three Vs [11]:

a) Volume – Large data sets (in terabytes) are collected from various sources including social media, financial transactions and sensor/machine registered information.
b) Velocity – Real time data collected with unprecedented speed including smart sensors and RFID tags that often require time-sensitive analysis.
c) Variety – Heterogeneous data collected in different forms or formats including email, video, audio and text documents.

Gaghmous and Kumar heralded the arrival of big data research that would bring forth novel data, methods and evaluation challenges to climate science that simultaneously require theory-guided data science methods to provide insightful climate data findings [12]. Accordingly, the sheer amount of data obtained from diverse sources could present significant challenges when one attempts to connect and correlate multiple data linkages. Labrinidis and Jagadish explained that the overall process involved in extracting insights from big data comprise two key processes namely data management (data storage and retrieval) and analytics (analysis and interpretation) [13].

IOP Conf. Series: Earth and Environmental Science **228** (2019) 012011 doi:10.1088/1755-1315/228/1/012011

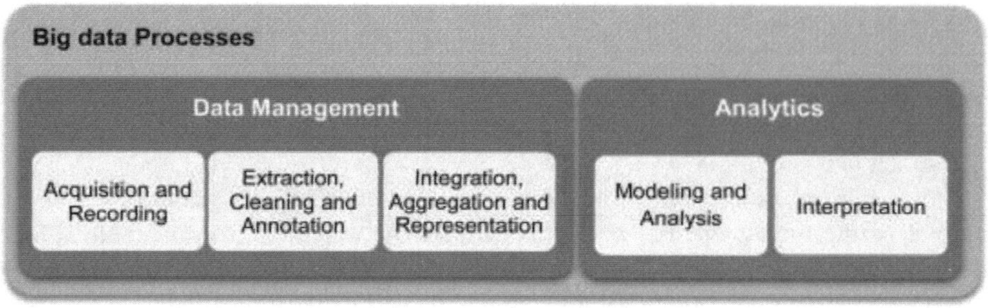

Figure 2. Big Data Processes.

The following table attempts to summarise the different data sets that fulfil the afore-mentioned 3Vs criteria and would require big data analytical tool-box to correlate climate change policy with the identified response measures:

Table 1. Data type/characteristics for identified parameters under climate response.

Response Measures	Data type/characteristics			Source of data
	Volume	Velocity	Variety	
Green Jobs				
Energy Consumption Cost	√	√	√	Energy bill encompassing the product life cycle from different energy sources.
Service/ Product Demand	√	√	√	Production Report
GHG emission	√	√	√	GHG Inventory , NRE
Wage Rate	√			Labour Dept. Statistics
Employment Numbers	√			Labour Dept. Statistics
Green Labelling				
Energy Consumption Cost	√	√	√	Energy bill encompassing the product life cycle from different renewable energy sources
Service/ Product Demand	√	√	√	Production Report
GHG emission	√	√	√	GHG Inventory , NRE
Employment Numbers	√			Labour Dept. Statistics
Cash Flow for Manufacturers	√			Sample financial statements from manufacturers
Cap-and-Trade Scheme				
GHG emission	√	√	√	GHG Inventory , NRE
Environmental preservation	√		√	Environmental Impact Assessment Report
Human Health Benefits	√	√	√	Health Statistics, MOH
Cost of compliance	√			Manufacturer's compliance record
Allowance Activity	√			

<u>Green Tariffs</u>				
Energy Consumption Cost	√	√	√	Energy bill encompassing the product life cycle from different energy sources
GHG emission	√	√	√	GHG Inventory , NRE

As illustrated above, many of the response measures share similar parameters and data sets particularly energy consumption cost, GHG emissions and service/product demand that fulfil the big data 3Vs criteria namely volume, variety, velocity. In this regard, big data analytics would allow stakeholders to track the vast troves of vital information such as production capacity and cost ('volume'), examine the energy consumption and GHG emission granularly ('variety') in real time ('velocity').

The establishment of a centralised data analytics structure to deposit the above shared data set would allow relevant stakeholders evaluate the overall impact of climate response implementation simultaneously on green jobs, green labelling and green tariffs. Moreover, a carefully constructed data analytic methodology including storage, scrubbing (transform information into readable dataset), profiling (end-to-end inventory of metadata), mapping (information flow), analysis (measuring of impact), visualisation and compliance of data would provide an outline of the sector's overall green performance in an aggregated manner [14]. By carefully studying the analytical results with in-depth understanding of the data linkages and interoperability, stakeholders will be able to evaluate the impact of response measures and adjust policy strategies in a fast changing climate environment.

3. Summary

The transformative impact of big data analytics is poised to alter the conventional data analysis by obtaining valuable insights concerning the correlation between climate risk and impact of the response measures implementation. By integrating a big data infrastructure with cross-border trade, industry performance, energy utilization and green job related database analytic modelling, the relevant stakeholders will be able to implement evidence-based climate policy decisions while limiting the adverse consequences arising from such measures. Aside from maintaining the momentum of sustainable development, countries will stand to benefit by developing better insights and positions while pursuing future climate change negotiations.

References

[1] International Centre for Trade & Sustainable Development 2011 *Biores* **5** P4
[2] Govindasamy P 2016 *Chinese Journal of Urban Environmental Studies* **05** P1
[3] International Labour Organization 2013 Methodologies for assessing green jobs *Geneva ILO*
[4] Bacon R and Kojima M 2011 Issues in estimating the employment generated by energy sector activities. *World Bank*
[5] Bowen A and Kuralbayeva K 2015 Looking for green jobs: the impact of green growth on employment. *Global Green Growth Institute*
[6] Dadhich P K 2017 Methodologies for impact assessment of implement response measures.
[7] Cohen A C and Vandenbergh M P 2012 The potential role of carbon labeling in a green economy. *Energy Economics.* **34** S53
[8] International Institute of Sustainable Development 2017 A guidebook on cross-border impact assessment of implemented response measures: subsidies, carbon trading and green tariffs
[9] United States Environmental Protection Agency 2003 Tools of the trade: a guide to designing

and operating a cap and trade program for pollution control. *US EPA*
[10] Belcher J and Sunkara R 2017 Green tariffs: attracting new loads with sustainable options
[11] Laney D 2001 3D data management: controlling data volume, velocity, and variety. *META*
[12] Faghmous J H and Kumar V 2014 *Big Data* **2(3)** 155
[13] Labrinidis A and Jagadish H V 2012 *VLDB Endowment Journal* **5** 2032
[14] Consoli R 2018 Using big data analytics to improve production. *manufacturing business tech. group Inc*

Energy Efficiency of Air Conditioners in Developing Countries: A Malaysian Case Study

Siti Fatihah Salleh[1,*], Aishah Mohd Isa[2], Mohd Eqwan Roslan[3], Tuan Ab Rashid Tuan Abdullah[1]

[1]Institute of Energy Policy and Research, Universiti Tenaga Nasional
[2]Department of Electrical Power Engineering, College of Engineering, Universiti Tenaga Nasional
[3]Department of Mechanical Engineering, College of Engineering, Universiti Tenaga Nasional
*Corresponding author's email: Siti.Fatihah@uniten.edu.my / sitifatihah.salleh@gmail.com

Abstract. Growing demand for air conditioner (AC) in developing countries has hit an alarming point. Enforcing higher efficiency standards for cooling appliances is one of the easier initiatives that governments can implement to reduce energy consumption and emissions of potent greenhouse gases. Malaysia has adopted the Minimum Energy Performance Standards (MEPS) in 2013 to regulate the energy performance of five common household appliances including ACs. This work sheds light on the overall improvement of AC's energy performance after MEPS implementation. A rigorous data collection exercise was conducted to build a comprehensive sales and technical database of MEPS-compliant ACs covering the years 2013-2015. In 2015, the average energy efficiency ratio (EER) of ACs sold in the local market is 3.4, and the cumulative electricity saving and CO_2 emission reduction after MEPS implementation are 1,764 GWh and 1,178 $ktCO_2eq$ respectively. The results from this study will contribute significantly to the review of the effectiveness of MEPS implementation in developing countries.

Keywords: energy efficiency; emissions reduction; minimum energy performance standards (MEPS); air conditioner; inverter.

1. Introduction

Air conditioner (AC) has become an essential comfort feature in modern homes, particularly for households in tropical and desert climates. Statistics have shown that the use of AC is escalating especially in developing countries in line with their rapid economic development and increased living standards [1]. Taking China for example, the adoption of room ACs in urban households had increased from less than a percent in 1990 to 112% in 2011, an astonishing increment of more than 300 times in just twenty years [2]. A survey conducted in several economically developed Chinese provinces also found that room AC ownership increases almost linearly with the increase of per capita annual income [3]. Many other countries echo similar circumstances [4]. **Figure 1** shows the growth of AC usage in 150 developing countries and emerging economies. The total number of stock AC in these countries is poised to hit 1.5 billion in 2030, which is about 94% of the amount of AC presently in use across the globe [5].

Content from this work may be used under the terms of the Creative Commons Attribution 3.0 licence. Any further distribution of this work must maintain attribution to the author(s) and the title of the work, journal citation and DOI.
Published under licence by IOP Publishing Ltd

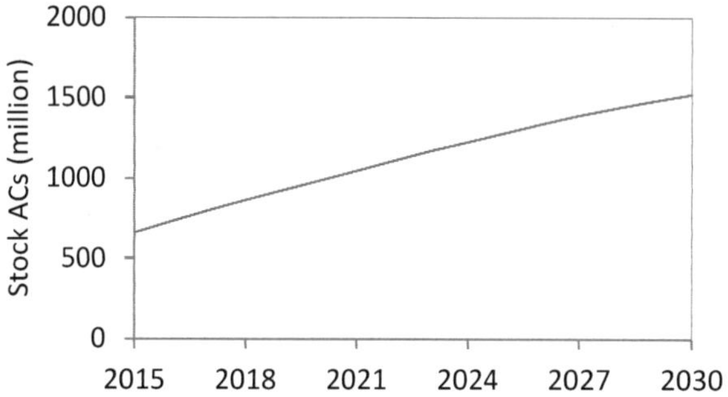

Figure 1: Projection of AC usage in developing countries (*Source: U4E Country Assessments*)

Modern AC nowadays does not merely cool the air, but is also equipped with nano-technology and smart air-purifying devices that are able to remove moisture and improve the indoor air quality by filtering airborne bacteria and viruses. Unfortunately, as good as it may sound, AC is also energy intensive; amounted to 1 trillion kilowatt hours of global energy consumption [1], therefore putting a strain on energy demand in many countries. Furthermore, the coolant or refrigerant widely used in AC, hydrofluorocarbons (HFCs), is a serious climate threat. A recent investigation by United Nation on 150 developing countries has found that AC alone was responsible for 640 million tonnes of carbon dioxide equivalent (CO_{2e}) emissions, which accounts for about 2% of the global energy-related CO_2 emissions in 2014 [6]. Under these circumstances, there is a growing emphasis on accelerating the adoption of more energy efficient and environmental friendly ACs.

If all developing countries adopted energy efficient products with proper standard settings and labelling, there would be massive energy saving and overall energy efficiency improvement of home appliance stock. A proper enforcement of energy efficiency standards is therefore necessary to increase penetration of high efficiency products into developing countries. The minimum energy performance standard (MEPS) is customarily used to specify the maximum amount of energy that a particular electrical device could consume in performing a certain task. MEPS has effectively brought significant economic and environmental benefits [5, 6] in more than 70 countries. In ASEAN countries, regulators and regional bodies are working diligently to advance their energy policies, strategies and actions to phase-out inefficient products especially AC and lighting.

2. AC Utilization in Malaysia

As a member of ASEAN countries, and one of the world's emerging economies, Malaysia is no exception to the rising AC usage. Malaysia has a tropical forest climate [7], with 13 out of its 35 cities identified as "mildly dry" to "dry"[8]. With rising economy and relatively warm temperature all year round, AC ownership in Malaysian household has increased from 31% in 2009 to 38% % in 2012 [9]. In 2013, the Energy Commission of Malaysia has implemented MEPS to standardize the energy performance of ACs. Standard and Industrial Research Institute of Malaysia (SIRIM) Berhad was tasked to develop the mandatory national standards. SIRIM has established *MS 2597:2014 Minimum Energy Performance Standards (MEPS) for Air Conditioner* which specifies the requirements for room ACs, classified as *"single-phase, non-ducted, single split wall mounted type vapour compression air conditioners with cooling capacity up to 7.1 kW*". Centralized and built-in ACs are currently excluded from the scope of the MS 2597:2014.

The Energy Efficiency Ratio (EER) and the Seasonal Energy Efficiency Ratio (SEER) are the two main types of metrics in use internationally to rate the energy efficiency of ACs. The EER metric is currently adopted in Malaysia to measure the ratio of the cooling capacity and the power consumed when measured at full load. Testing may be done locally at SIRIM or abroad at test laboratories recognized by the Department of Standards Malaysia. Based on the tested EER, star rating will be assigned to the product model in accordance to MS 2597:2014.The MEPS star rating for the rated cooling capacity (RCC) are tabulated in **Table 1**.

Table 1: MEPS Star Rating for AC

(i) RCC< 4.5 kW

Star Rating	Tested EER (BTu/h)/W
5	>11.94
4	11.16 -11.93
3	10.37 - 11.15
2	9.56 - 10.36
1	9.00 - 9.55

(ii) 4.5 kW ≤ RCC ≤7.1 kW

Star Rating	Tested EER (BTu/h)/W
5	>10.71
4	9.83 - 10.70
3	8.94 - 9.82
2	8.03 - 8.93
1	7.50 - 8.02

Before MEPS, there is no established performance monitoring and quantification of the EER for ACs in Malaysia. Quantifying how much energy have been saved from MEPS program is of increasing national interest, especially with increasing environmental concerns and rising energy demand. This need has led the authors to conduct a study to assess the effectiveness of MEPS implementation in Malaysia.

3. Methodology

3.1. Data Collection Method

The authors conducted a rigorous data collection of MEPS-compliant AC sales covering the years 2013-2015. The sales data were acquired from major manufacturers and distributors that hold Certificate of Approval (COA) by the Energy Commission, with 95% coverage of the data set of all MEPS-compliant appliances (1,150 COAs out of the 1, 215 COAs issued by 2015). Technical data on cooling capacity and power input were also collected and all information consolidated into a comprehensive database.

3.2. Energy Efficiency Ratio (EER)

The EER is the metric adopted in Malaysia to assign energy efficiency rating for MEPS-compliant AC.

$$EER = \frac{Cooling\ Capacity\ (W)}{Power\ Input\ (W)} \tag{1}$$

Where 1 W = 3.412 Btu/hr

3.3. Annual Electricity Consumption (AEC)

To calculate annual electricity consumption for each AC model, the following equations are used as recommended by SIRIM for both AC types assuming 8 hours running time per day;

For non-inverter type,

$$A = \frac{365\ days \times 8\ \frac{h}{day} \times Power\ Input\ measured\ (W)}{1000} \tag{2}$$

For inverter type,

$$A = \frac{365\ days \times 8\ \frac{h}{day} \times ((0.4 \times Power\ Input\ _{full-load}, W) + (0.6 \times Power\ Input\ _{half-load}, W))}{1000} \tag{3}$$

Where A [kWh] is the annual electricity consumption for each AC model.

The power input $_{half\text{-}load}$ and the tested cooling capacity (Btu/hr) are reported in SIRIM's test report. To calculate total annual electricity consumption, the total annual electricity consumption of ACs determined previously was multiplied by the number of models sold in that year.

$$AEC_{total,year} = \sum(AEC_{model} \times NUS_{model,year}) \tag{4}$$

Where $AEC_{total,year}$ [kWh] is the total electricity consumed by MEPS-compliant ACs in a particular year while AEC_{model} [kWh] is the annual electricity consumption for each model $NUS_{model,year}$ is the number of units sold for a model in that same year.

3.4. Annual Electricity Saving (AES)

The maximum potential electricity saving was estimated by comparing the calculated annual electricity consumption with the annual consumption of the lowest 2-star rating model of the same equipment type, a similar approach applied by Energy Commission in the National Energy Efficiency Action Plan (NEEAP) [5]. By using the tested cooling capacity and EER as reference, the electricity consumption per year for the lowest 2-stars rating model measured (kWh) can be calculated by using Equation 5 and 6 below.

i) For cooling capacity of $\leq$ 4.5kW, $EER_{Lowest\ 2\text{-}star\ model} = 2.80$,

$$B = 365\ days \times 8\ \frac{h}{day} \times \frac{Tested\ cooling\ capacity\ (kW)}{2.80} \tag{5}$$

ii) For cooling capacity of 4.5kW $\leq$ cooling capacity $\leq$ 7.1kW , $EER_{Lowest\ 2\text{-}star\ model} = 2.35$,

$$B = 365\ days \times 8\ \frac{h}{day} \times \frac{Tested\ cooling\ capacity\ (kW)}{2.35} \tag{6}$$

Where *B* [kWh] is the annual electricity consumption for each AC model in reference to 2-stars rating model.

The amount of energy savings compared to the lowest 2-stars rating model was then calculated using equation 7 given below;

$$\text{Energy savings per model} = A - B \tag{7}$$

Thus, the total annual energy saving, gained for both AC types can be estimated using equation 8 as below;

$$AES_{total,year} = \sum(AES_{model} \times NUS_{model,year}) \tag{8}$$

Where $AES_{total,year}$ [kWh] is the total electricity saving by MEPS-compliant ACs in a particular year while AES_{model} [kWh] is the annual electricity saving for each model and $NUS_{model,year}$ is the number of units sold of a model in that same year.

3.5. CO_2 Emission Reduction (CER)

As for the annual carbon emission reduction, $CER_{total,year}$ [ktCO_2eq], it is determined based on the total annual electricity saving, $AES_{total,year}$ [GWh] times by the grid emission factor, GEF_{region} [tCO_2eq/MWh].

$$CER_{total,year} = AES_{total,year} \times GEF_{region,year} \times GA \tag{9}$$

Where *GA* [%] is the grid allocation.

Malaysia is served by three electricity grids; the TNB grid for Peninsular Malaysia that serves about 90% of the electricity demand, the SEB grid for Sarawak and the SESB grid for Sabah. Sabah and Sarawak are states on the East Coast of Malaysia. Since each grid has different profiles for generation capacity, the grid emission factor is also different for each grid as tabulated listed in **Table 2**.

Table 2: Grid Emission Factor by Region

Grid emission factor (tCO_2/MWh) **	2013	2014	2015
Peninsular Malaysia (TNB Grid)	0.742	0.694	0.659
Sabah (SESB Grid)	0.533	0.536	0.572
Sarawak (SEB Grid)	0.430	0.335	0.421

Source: Compiled from [10]; [11];[12]

The annual carbon emissions reductions are allocated amongst the three grids using the same ratio adopted in NEEAP, whereby 90% of the carbon emissions reductions is allocated to Peninsular Malaysia, while Sabah and Sarawak are allocated 5% each.

4. Results and Discussion

4.1. Energy Efficiency Performance

Figure 2 indicates the energy performance of ACs sold in the local market from year 2013 to 2015. The average energy efficiency measured as the Energy Efficiency Ratio (EER) increases very slightly from 3.34 in 2013 to 3.40 in 2915. The average annual efficiency increment is close to 1%. However, the range of available EER sold in the market does show an interesting yearly improvement. The MEPS implementation has raised the bar for minimum EER by 8% from 2.38 in 2013 to 2.56 in 2015. In the meantime, the maximum EER shows higher improvement with a dramatic increase from 4.57 in 2013 to 5.24 in 2015, constituting a 14% increment. The Malaysian market average of EER is found to be comparable to India which has average EER range between 3.0 – 4.0 [6]. It is also slightly higher than China's minimum efficiency for AC which is 3.2 in 2009 [1]. Since China plays a major role as the world's top AC manufacturer and exporter, China's minimum EER can be regarded as the current best practice.

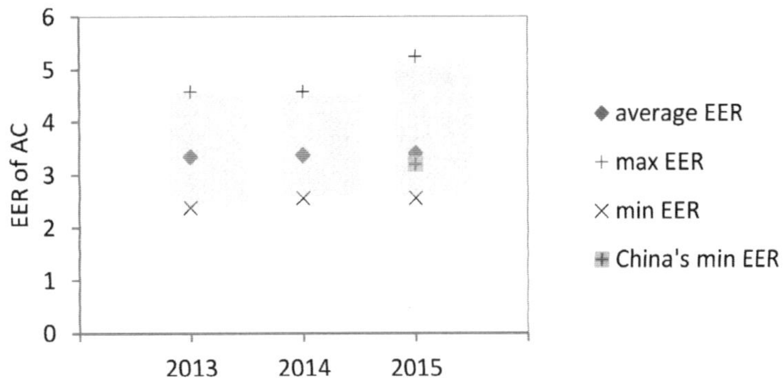

Figure 2: Energy performance of ACs after MEPS implementation.

4.2. Annual Electricity Consumption and Saving

The annual electricity consumption and saving by MEPS-compliant ACs are shown in **Figure 3**. Generally, both of them increase every year with average annual increment of 12% and 38%

respectively. Higher rate of annual electricity saving can be observed as it leaps from 404 GWh in 2013 to 762 GWh in 2015, caused by increased adoption of more efficient ACs especially the inverter type. Based on our anlalysis, inverter type AC consumes about 30-40% less electricity than the non-inverters. Inverter type AC in general has the flexibility to vary the compressor's speed and power in accordance to the cooling requirement of the room. On the other hand, regular type AC will have to run at full capacity and peak power requirement every time the compressor is running hence consuming more energy. In order to further improve AC's energy efficiency, innovative technologies could also be applied to fans, heat exchangers, expansion devices, and refrigerant fluids. The potential saving that could be gained if all these technologies were applied together in an AC unit is between 60 – 72 per cent of energy compared to non-inverter model [6].

Meanwhile, the annual electricity consumption of ACs also steadily increases every year reaching 3,128 GWh in 2015. Even though the government has attempted to reduce energy consumption per appliance through MEPS, the total energy consumption still increases every year due to increased sales volume of ACs. The average sales growth of AC from 2013 to 2015 is about 13% annually, most likely due to increased population and per capita annual income. Besides that, improved efficiency could also influence people to buy more AC unit for their house since the efficient models consume less electricity hence saves money, producing what is known as the rebound effect.

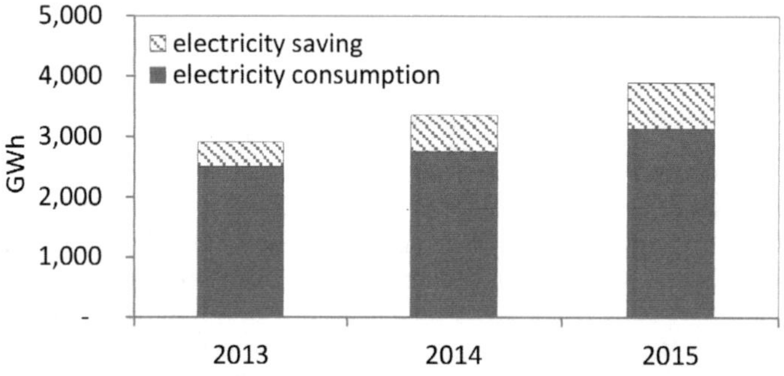

Figure 3: Annual electricity consumption and saving of MEPS-compliant ACs.

4.3. CO_2 Emission Reduction

Figure 4 shows the resulting annual emissions reductions from the increase of AC's energy efficiency. Assuming that emission reduction will carry over annually after the purchase of ACs in a given year, the total cumulative amount of emissions reduction after MEPS implementation on AC in 2015 is 1,178 $ktCO_2eq$.

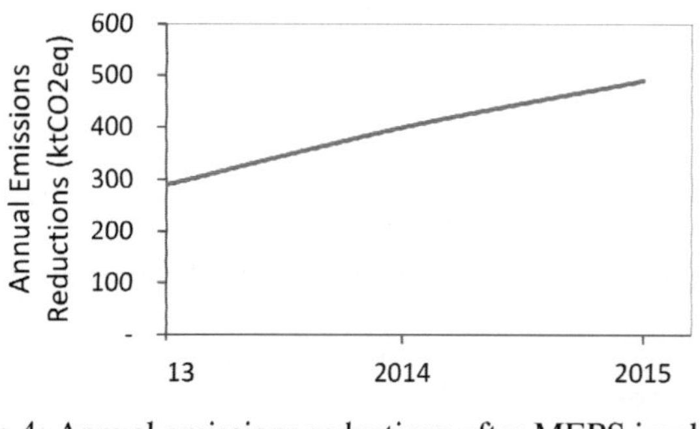

Figure 4: Annual emissions reductions after MEPS implementation.

5. Conclusion and Future Works

Increased energy use has become a critical global issue. Setting higher energy efficiency standards for household electrical appliances has been proven to reduce the rising energy demand. The increase of the market average of EER proves that MEPS has increased the diffusion of efficient ACs into Malaysian market. EER is typically the metric chosen when MEPS is being implemented in a country for the first time. However, many developed countries including Singapore has already transitioned to the Seasonal Energy Efficiency Ratio (SEER) metric which is actually a derivative of EER with more complicated formula to weigh in the different part load periods and variable speed systems. Malaysia could adopt SEER for AC with variable speed system in the future as it is more representative of the seasonal performance of AC, and also better at estimating the potential energy savings. Besides SEER, since Malaysia falls into the tropical climate zone, the Cooling Seasonal Performance Factor (CSPF) metric could also be adopted as it takes into account the ratio between the annual amount of heat removed and the amount of energy consumed. So far, CSPF has been adopted by Republic of South Korea and Vietnam.

As one of developing countries with manufacturing capacity, Malaysia is already a member of ASEAN SHINE, which is a platform for harmonizing energy efficiency initiatives in ASEAN. Harmonization of the energy efficiency standards and testing method and should be able to path the way to an integrated regional policy approach so that the diffusion of highly efficient products could be accelerated and sustained by the domestic market. Focusing on ASEAN SHINE initiatives for ACs, ASEAN member countries have agreed to a uniform testing method derived from ISO5151-2010 and to a minimum EER of 2.9W/W or CSPF of 3.08W/W by 2020 as mandatory MEPS for all fixed and variable drive ACs below 3.52kWcapacities.

The large annual increment of total energy consumption of ACs also alludes to a rebound effect that may counteract the MEPS policy measures to reduce energy consumption in the residential sector. Rebound effect occurs when behavioural or a systemic response negates the savings achieved through technological improvements. As AC prices are likely to decrease due to economies of scale, the ownership of AC would accelerate even more. As a consequence, the energy saving target could not be achieved. For that reason, complementary efforts to educate consumers on smart purchasing and smart behavior when utilizing ACs are also important to minimize rebound effect.

Acknowledgement

This work was supported by the Energy Commission through the Chair in Energy Economics of Energy Commission at IEPRe UNITEN research fund [grant number KETST2016002].
We would like to express our heartfelt gratitude to the Energy Commission, the Chair of IEPRe, Prof. Dr. Ken Koyama, representatives from Malaysian Air-Conditioning & Refrigeration Association (MACRA), Malaysian Electrical Appliances Distributions Association (MEADA) as well the workshop participants and survey respondents.

References

[1] M. a. McNeil and V. E. Letschert, "Future air conditioning energy consumption in developing countries and what can be done about it: the potential of effi ciency in the residential sector," 2008.

[2] J. Liu, X. Sun, B. Lu, Y. Zhang, and R. Sun, "The life cycle rebound effect of air-conditioner consumption in China," *Appl. Energy*, vol. 184, pp. 1026–1032, 2016.

[3] W. Long, T. Zhong, and B. Zhang, "China: The Issue of Residential Air Conditioning," 2018.
[4] L. W. Davis and P. J. Gertler, "Contribution of air conditioning adoption to future energy use under global warming," *Proc. Natl. Acad. Sci.*, vol. 112, no. 19, p. 5962 LP-5967, May 2015.
[5] OECD/IEA, "Future of cooling: Opportunities for energy-efficient air conditioning," 2018.
[6] UN Environment-Global Environment Facility, "Accelerating the Global Adoption of Energy-Efficient and Climate-Friendly Air Conditioners," p. 97, 2017.
[7] M. Peel, B. Finlayson, and T. McMahon, "Updated world map of the Köppen-Geiger climate classification," *Hydrol. Earth Syst. Sci.*, vol. 11, pp. 1633–1644, 2007.
[8] Energy Commission (Malaysia), "Energy Efficiency Criteria for Electrical Equipment to Qualify for the Minimum Energy Performance Standards Star Rating," 2018. [Online]. Available: Next on the line. [Accessed: 23-Aug-2018].
[9] Department of Statistic (Malaysia), "Statistic Data Warehouse (e-Data Bank)," 2018. [Online]. Available: https://www.dosm.gov.my/v1/index.php?r=column/cthree&menu_id=cEhBV0xzWll6WTRjdkJienhoR290QT09 .
[10] Malaysia Green Technology Corporation, "Study on Grid Connected Electricity Baselines in Malaysia (Year 2012, 2013 & 2014)," 2016. [Online]. Available: http://www.seda.gov.my/co2_avoidance.html . [Accessed: 04-Apr-2018].
[11] Ministry of Natural Resources and Environment (Malaysia), "Malaysia Second Biennial Update (BUR) to United Nations Framework Convention on Climate Change (UNFCC)-Final Draft Version," 2018.
[12] Sarawak Energy, "Sustainability Report 2015," 2016.

IOP Conf. Series: Earth and Environmental Science **228** (2019) 012014 doi:10.1088/1755-1315/228/1/012014

Disposable bismuth-based electrodes for heavy metal ion detection

Sandeep Gupta[1*], KM. Sapna[2], Kanad Ray[3*], S L Kothari[1], Kamlendra Awasthi[2] and Manoj Kumar[2]

[1] Amity Institute of Biotechnology, Amity University Rajasthan, Kant Kalwar, NH-11C, Jaipur-303007, Rajasthan, India

[2] Department of Physics, Malaviya National Institute of Technology, JLN Marg, Jaipur-302017, Rajasthan, India

[3] Amity Institute of Applied Sciences, Amity University Rajasthan, Kant Kalwar, NH-11C, Jaipur-303007, Rajasthan, India

E-mail: sgupta.phy@gmail.com,kray@jpr.amity.edu

Abstract. This work presents the efficacy of bismuth-based mixed oxide nanosructures for heavy metal ion detection. Four samples with varying proportions of Na_3BiO_4, Bi_2O_3 and Na_2O were synthesized using potentiostatic electrodeposition. X-ray diffraction (XRD) indicates the presence of poly-crystalline Na_3BiO_4 and Na_2O in sample 1 while presence of poly-crystalline Na_3BiO_4 and Bi_2O_3 were seen in samples 2 and 3. Poly-crystalline Bi_2O_3 was seen in sample 4. Scanning electron microscopy (SEM) showed the presence of microplates of varying shapes and sizes with an average thickness $< 1\mu m$. Linear stripping voltammetry confirms that Sample 2 shows highest sensitivity towards detection of heavy metal ions.

1. Introduction

Heavy metal ion detection is quite significant from the environmental point of view [1]. Dumping of industrial waste into rivers and oceans has led to their contamination with heavy metal ions like lead and mercury [2, 3]. Even trace concentrations of these metals can severely affect the health of living beings [4, 5]. Developing countries suffer most due to this contamination as proper diagnostic tools are not available in remote areas. If cheap, disposable and portable sensors for detection of these heavy metals in water can be developed, than on-site analysis of drinking water samples is possible in the remote areas. Electrochemical sensors provide an ideal way of achieving this [6]. Stripping voltammetry is one of the most promising electrochemical techniques for high sensitivity and remote analysis [7, 8]. However their efficiency and effectiveness is dependent on the material used at the working electrode. Development of new cost-effective and efficient materials for such sensors requires global research efforts.

Bismuth is an unique element and finds application is areas like hydrogen storage [9, 10], detection of heavy metal ions [11], photocatalysis [12, 13], electrocatalysis [14, 15] etc. Bismuth-based electrodes are gaining wide popularity for heavy metal ion sensing because of its low toxicity, partial insensitivity to dissolved oxygen, ability to form a number of alloys [16]. Many potential bismuth-based alloys can be developed for application in heavy metal ion detection [17, 18, 19, 20, 21, 22, 23] .

Content from this work may be used under the terms of the Creative Commons Attribution 3.0 licence. Any further distribution of this work must maintain attribution to the author(s) and the title of the work, journal citation and DOI.
Published under licence by IOP Publishing Ltd

In this work, four samples were prepared using electrodeposition. Sample 1 was prepared by using Na^{+} and Bi^{3+} ions in a molar ratio of 5:1. The molar ratio was changed to 3:1 for sample 2. Sample 3 has a molar ratio of 1:1 while sample 4 was only prepared by using Bi^{3+} ions. These samples were then tested for their sensitivity towards detection of Pb^{2+} and Hg^{2+} using linear stripping voltammetry.

2. Experimental Section

2.1. Materials

Bismuth nitrate pentahydrate ($Bi(NO_3)_3.5H_2O$) was purchased from Loba. Nitric acid (HNO_3) was purchased from Qualigens while sodium nitrate ($NaNO_3$) was purchased from Merck. Mercury (II) Chloride ($HgCl_2$) and Lead (II) Chloride ($PbCl_2$) were purchased from Sigma Aldrich.

2.2. Synthesis

Electrodeposition with three electrode setup was used for synthesis. ITO substrate was used as the working electrode while platinum wire was used as the counter electrode. Ag/AgCl in saturated KCl was used as the reference electrode [24, 25]. Four different electrolytes were prepared. First electrolyte contained $NaNO_3$, $Bi(NO_3)_3.5H_2O$ and HNO_3 in molar concentrations of 0.065 M, 0.013 M and 1 M respectively. Second electrolyte had molar concentrations of 0.039 M, 0.013 M and 1 M respectively. Third electrolyte was prepared by adding the respective salts in the molar ratio of 0.013 M, 0.013 M and 1 M respectively. Fourth electrolyte contained only $Bi(NO_3)_3.5H_2O$ and HNO_3 in the molar concentrations of 0.013 M and 1 M respectively. Deposition was performed at a reduction potential of -0.15 V. Solutions containing 1 ppm each of Hg^{2+} and Pb^{2+} ions was prepared by using $HgCl_2$ and $PbCl_2$ for heavy metal ion detection.

2.3. Characterization

Structural studies were performed using XRD (Panalytical X'pert Pro). Morphology was studied using SEM (Nova nanoSEM FEI). Detection of heavy metal ions was done using potentiostat purchased from Kanopy Techno Solutions Pvt. Ltd., Indian Institute of Technology, Kanpur, India.

3. Results and discussion

3.1. Morphological studies

Four samples were prepared by varying the concentration of sodium. Scanning electron microscopy was performed to understand the morphology of the synthesized samples (figure 1). Sample 1 showed the presence of microplates with an average thickness of 600 nm. Regular hexagons with an average thickness of 500 nm were seen in sample 2. Sample 3 again shows micro plates with thickness varying from 400 nm to 600 nm. Sample 4 shows much thicker microplates (Average thickness 800 nm).

3.2. Structural studies

Samples were studied for the structural properties using XRD (figure 2). The results indicate the formation of crystalline sodium bismuth oxide (Na_3BiO_4), bismuth oxide (Bi_2O_3) and sodium oxide (Na_2O) in different proportions (JCPDS:01-071-1583, 00-041-1449, 01-077-2148 respectively). Table 1 summarizes the composition of all the samples.

Figure 1. Scanning electron microscopy images for Sample 1 (a), Sample 2 (b), Sample 3 (c) and Sample 4 (d)

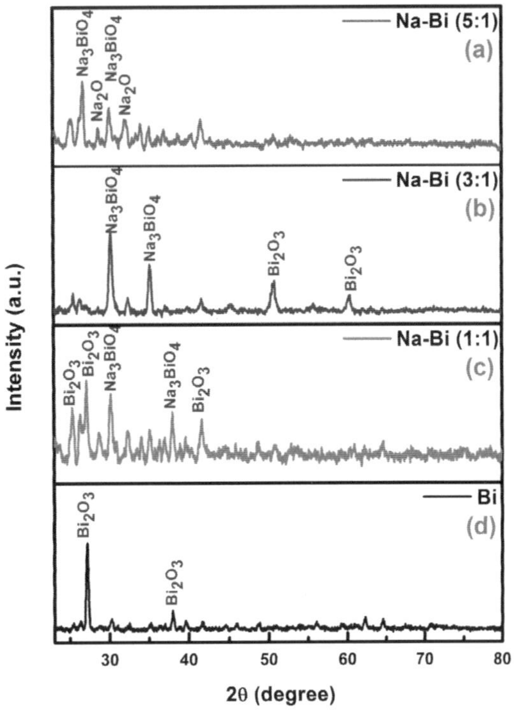

Figure 2. X-ray diffraction pattern for Sample 1 (a), Sample 2 (b), Sample 3 (c) and Sample 4 (d)

Table 1. Percentage composition of Sample 1, Sample 2, Sample 3 and Sample 4

Sample	Na_3BiO_4 (%)	Bi_2O_3 (%)	Na_2O (%)
1.	70%	0%	30%
2.	50%	50%	0%
3.	20%	80%	0%
4.	0%	100%	0%

3.3. Heavy metal ion detection

Heavy metal ion detection was performed using linear stripping voltammetry (figure 3). Result shows two separate peaks at -0.25 V and 0.85 V respectively. These peaks correspond to Pb^{2+} and Hg^{2+} ions respectively. Peak corresponding to Hg^{2+} ions has lowest intensity for sample 1. For samples 2 to 4, a gradual decrease in its intensity is observed. SEM results show the presence of microplates with thickness varying from 500 nm to 800 nm. The percentage composition of various oxides is mainly responsible for this phenomenon. As seen in Table 1, a gradual decrease in the percentage of Na_3BiO_4 is seen from Sample 2 to 4. This observation is in line with the decreasing peak intensity. Thus it can be concluded that higher percentage of Na_3BiO_4 gives better results towards the detection of Hg^{2+} ions. Sample 1 gives lowest peak intensity. Although it has highest percentage of Na_3BiO_4, it shows lowest sensitivity. This is probably due to the replacement of Bi_2O_3 by Na_2O. Thus it appears that a mixed oxide comprising of Na_3BiO_4 and Bi_2O_3 is ideally suited for heavy metal ion detection. Similar results are seen for Pb^{2+} ions with minor variations probably due to variation of concentration around the working electrode.

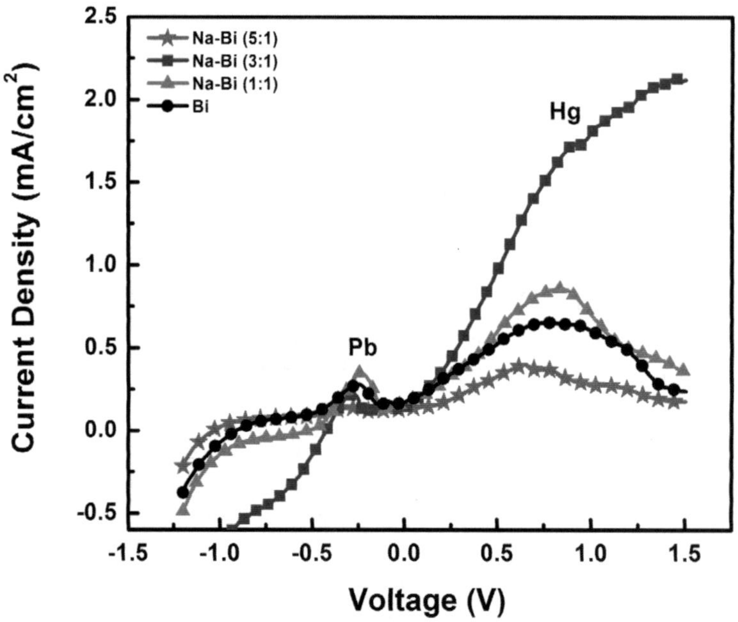

Figure 3. Heavy metal ion detection (Hg^{2+} and Pb^{2+}) using Sample 1 to 4

Other bismuth-based elctrodes reported for heavy metal ion detection are summarized in Table 2. 3D honeycomb-like N-doped carbon nanosheet framework decorated with bismuth nanoparticles [17], Graphene-cysteine composite modified bismuth film electrode [18] and glassy carbon electrode modified with Bi, nafion and 2-mercaptoethanesulfonate (MES)-tethered polyaniline (PANI) [19] shows good sensitivity.

Na_3BiO_4 and Bi_2O_3 mixed oxide electrodes offer great potential for application in heavy metal ion detection. These electrodes are disposable and easy to synthesize. The average cost associated with the synthesis of a single electrode is much below the industry average. The toxicity is also very less as compared to the traditionally used mercury based electrodes. Nanostructures further enhance the sensitivity due to the availability of large surface area [26]. Thus by tuning the shape and size of the nanostructures, the sensitivity can be tuned for specific metal ion.

Table 2. Various bismuth-based alloys for heavy metal ion detection

Electrode	Reference
Bi/N doped Carbon	[17]
Bi/Gold/Graphene/Cysteine	[18]
Bi/Nafion/PANI	[19]
Bi/Carbon Paste	[20]
Bi/Zeolite/Carbon Paste	[21]
Bi/Nafion/Medical Stone	[22]
BiOCl/MWCNT/GCE	[23]

4. Conclusions

Mixed oxide nanostructures with different proportions of Na_3BiO_4, Bi_2O_3 and Na_2O were synthesized using potentiostatic electrodeposition. XRD indicated the presence of polycrystalline metal oxides. SEM revealed the presence of microplates with average thickness of 650 nm. Sample 2 showed best performance towards detection of heavy metal ions. A gradual decrease in sensitivity is seen with decreasing proportion of Na_3BiO_4. Sample 1 with no Bi_2O_3 showed lowest sensitivity.

Acknowledgments

We thank DST, New Delhi (IFA-13/PH-84), SERB, New Delhi (ECR/2016/1888) and UGC DAE CSR, Indore (CSR-IC/CRS-73/2014-15/581) for providing financial support. We are also thankful to MNIT Jaipur for providing characterization facilities.

References

[1] Zou Y, Wang X, Khan A, Wang P, Liu Y, Alsaedi A, Hayat T and Wang X 2016 *Environmental science & technology* **50** 7290–7304
[2] Bagheri H, Afkhami A, Khoshsafar H, Rezaei M and Shirzadmehr A 2013 *Sensors and Actuators B: Chemical* **186** 451–460
[3] Privett B J, Shin J H and Schoenfisch M H 2010 *Analytical Chemistry* **82** 4723–4741
[4] Chow E, Hibbert D B and Gooding J J 2005 *Analytica Chimica Acta* **543** 167–176
[5] Li C L, Liu K T, Lin Y W and Chang H T 2010 *Analytical Chemistry* **83** 225–230
[6] Zhang B, Chen J, Zhu H, Yang T, Zou M, Zhang M and Du M 2016 *Electrochimica Acta* **196** 422–430
[7] Lu Y, Liang X, Niyungeko C, Zhou J, Xu J and Tian G 2018 *Talanta* **178** 324–338
[8] Cui L, Wu J and Ju H 2015 *Biosensors and Bioelectronics* **63** 276–286
[9] Gupta S, Singh R, Agarwal G, Ray K, Kothari S, Verma A S, Saraswat V K, Awasthi K and Kumar M 2018 *International Journal of Hydrogen Energy*

[10] Ni J, Bi X, Jiang Y, Li L and Lu J 2017 *Nano Energy* **34** 356–366
[11] Gupta S, Singh R, MD A, Kulshrestha V, Srivastava D N, Ray K, Kothari S, Awasthi K and Kumar M *Unpublished*
[12] Selvamani T, Anandan S, Granone L, Bahnemann D W and Ashokkumar M 2018 *Materials Chemistry Frontiers* **2** 1664–1673
[13] Jiang G, Li X, Lan M, Shen T, Lv X, Dong F and Zhang S 2017 *Applied Catalysis B: Environmental* **205** 532–540
[14] Han N, Wang Y, Yang H, Deng J, Wu J, Li Y and Li Y 2018 *Nature communications* **9** 1320
[15] Lv W, Zhou J, Bei J, Zhang R, Wang L, Xu Q and Wang W 2017 *Applied Surface Science* **393** 191–196
[16] Kokkinos C and Economou A 2008 *Current Analytical Chemistry* **4** 183–190
[17] Lu Z, Dai W, Lin X, Liu B, Zhang J, Ye J and Ye J 2018 *Electrochimica Acta* **266** 94–102
[18] Zhu L, Xu L, Huang B, Jia N, Tan L and Yao S 2014 *Electrochimica Acta* **115** 471–477
[19] Chen L, Su Z, He X, Liu Y, Qin C, Zhou Y, Li Z, Wang L, Xie Q and Yao S 2012 *Electrochemistry Communications* **15** 34–37
[20] Hočevar S B, Švancara I, Vytřas K and Ogorevc B 2005 *Electrochimica Acta* **51** 706–710
[21] Cao L, Jia J and Wang Z 2008 *Electrochimica Acta* **53** 2177–2182
[22] Li H, Li J, Yang Z, Xu Q, Hou C, Peng J and Hu X 2011 *Journal of hazardous materials* **191** 26–31
[23] Song Y and Chen J 2018
[24] Wolcott A, Smith W A, Kuykendall T R, Zhao Y and Zhang J Z 2009 *Small* **5** 104–111
[25] Durmus A, Gunbas G E, Camurlu P and Toppare L 2007 *Chemical Communications* 3246–3248
[26] Kang X, Wang J, Wu H, Aksay I A, Liu J and Lin Y 2009 *Biosensors and Bioelectronics* **25** 901–905

Exposure to $PM_{2.5}$ in urban area and respiratory health symptoms among urban workers in Klang Valley

Mohd Fairus Awang[1, 2], Juliana Jalaludin[1], Mohd Talib Latif[3], Noor Fatihah Mohamad Fandi[1]

[1]Department of Environmental and Occupational Health, Faculty of Medicine and Health Sciences, Universiti Putra Malaysia, 43400 UPM, Serdang, Selangor, Malaysia.
[2]Faculty of Engineering and Built Environment, Universiti Kebangsaan Malaysia, 43600 UKM, Bangi, Selangor, Malaysia.
[3]School of Environmental and Natural Resource Sciences, Universiti Kebangsaan Malaysia, 43600 UKM, Bangi, Selangor, Malaysia

E-mail: juliana@upm.edu.my

Abstract. Urban workers are susceptible to traffic-related air pollution (TRAP) exposure that can increase the prevalence of respiratory health symptoms among them. $PM_{2.5}$ was the main pollutants emitted from vehicle emission during fossil fuel combustion. This will influence TRAP level in an urban environment. This study was a cross-sectional comparative study that conducted among traffic policemen (n = 160) and office workers (n = 149) in Klang Valley. The respondents' information and respiratory health symptoms were gathered using a set of questionnaire that adapted from American Thoracic Society (ATS). The air samples for personal exposure and workplace exposure were collected during working hours. The average $PM_{2.5}$ levels for traffic policemen was 87μg/m^3 (personal) and 86μg/m^3 (workplace). The average $PM_{2.5}$ levels for office workers was 28μg/m^3 (personal) and 26μg/m^3 (workplace). The statistical results of PM2.5 levels between both groups are significant difference for personal exposure ($z = -15.685$, $p < 0.001$) and workplace exposure ($z = -14.519$, $p < 0.001$). The Chi-Square test result indicates that there is a significant difference in cough episode between both groups. This study concludes that exposure to high levels of $PM_{2.5}$ from TRAP would increase the prevalence of respiratory health symptoms among urban workers.

1. Introduction

Human activities such as transportation activity, burning solid fuel, and open burning contribute pollutants to urban air [1]. Traffic-related air pollutant (TRAP) was a major contributor to air pollution in urban areas [2] whereas vehicle emissions are the main source of TRAP [3]. According to the International Agency for Research on Cancer (IARC), outdoor air pollution was classified as a group 1 carcinogen to human. Furthermore, Fanizza et al., (2018) have found that vehicular exhaust was the main source of PM with aerodynamic diameter 2.5 μm ($PM_{2.5}$) based on the physicochemical [4]. In addition, the PM from TRAP consists of multiple chemical compounds, such as polycyclic aromatic hydrocarbon (PAH) [3, 5, 6]. The study finding by Silva et al. (2015) found that PAH can cause DNA damage and acts as an endpoint impact of air pollution [7].

Content from this work may be used under the terms of the Creative Commons Attribution 3.0 licence. Any further distribution of this work must maintain attribution to the author(s) and the title of the work, journal citation and DOI.
Published under licence by IOP Publishing Ltd

TRAP not just deteriorates air quality in an urban environment, but also influences health risk among urban dwellers. Patton et al., (2014) suggested that residents that live near a busy road were likely to have asthma, cardiovascular and respiratory diseases, and a decreased lung function [8]. A report from Abidin et al., (2014) study on exposure of TRAP among schoolchildren in Selangor at age 10 to 11 years old found that exposure to TRAP was significantly associated with asthma symptoms among schoolchildren [9]. Meanwhile, according to Bowatte et al., (2018), TRAP exposure also has a strong association with both the incidence and persistence of asthma in middle age adults [10].

Klang Valley that is located at the center of Selangor also is a fast-growing and highly populated area in Malaysia and well known for the development of industries and community activities. The Road Transport Department Malaysia (JPJ) reported about four percent increment of vehicles on the road in Malaysia (based on reports 2012 to 2014). Therefore, this will promote the anthropogenic generation of air pollution and increase health risk in the community especially those who work outdoors. Urban workers such as traffic policemen are susceptible exposure to high levels of TRAP. Therefore, they are at the highest risk of health effect compared to those who work indoor [11]. The aim of this study is to compare the $PM_{2.5}$ level between outdoor and indoor workers. In addition, the present study also compares the prevalence of respiratory health symptoms among respondents in Klang Valley.

2. Materials and methods

The cross-sectional comparative study was conducted between 160 traffic policemen (exposed group) and 149 office workers (comparative group) in Klang Valley. The respondents were selected based on gender, employment year and health exposure. They were male of between 20 to 60 years old, more than one-year employment, and never had radiotherapy, chemotherapy, therapeutic drug use or exposure to X-ray diagnosis in the past six months. This study was also approved by the Ethical Committee of Universiti Putra Malaysia (JKEUPM-2017-247). To gather the information on socio-demographic information and respiratory health status a set of questionnaire adapted from American Thoracic Society (ATS) was distributed among respondents. The air samples were collected during the working period for both personal and workplace exposure. $PM_{2.5}$ was measured by using Sidepak AM520 Personal Aerosol Monitor (TSI) for personal exposure and Dusttrak Aerosol Monitor 8534 (TSI) for workplace exposure. The data was analyzed using Statistical Package for Science Version 23 (SPSS Ver. 23) for statistical analysis. The comparison of $PM_{2.5}$ between traffic policemen and office workers was analyzed using the Mann-Whitney U test. Meanwhile, Chi-Square test was conducted for the comparison of respiratory health symptoms between both groups.

3. Results and discussions

3.1. Comparison of $PM_{2.5}$ exposure between exposed and comparative group

195 respondents were selected for personal air sampling. Figure 1 shows the average concentrations of $PM_{2.5}$ between traffic policemen and office workers at the workplace and personal exposure. The average personal exposure of $PM_{2.5}$ was 87µg/m^3 for traffic policemen and 28µg/m^3 for office workers. The average of the $PM_{2.5}$ level at the workplace for traffic policemen was 86µg/m^3 and for office workers was 26µg/m^3. Mann - Whitney U test was conducted to compare pollutants exposure among both groups. Table 1 shows the comparison of $PM_{2.5}$ exposure between both groups. There is a significant difference of $PM_{2.5}$ exposure between traffic policemen and office workers for personal exposure ($z = -15.685$, $p < 0.001$) and workplace exposure ($z = -14.519$, $p < 0.001$. The present study indicates that the traffic policemen experience $PM_{2.5}$ level above the guideline limits according to WHO and Malaysia standard for 24 hours exposure.

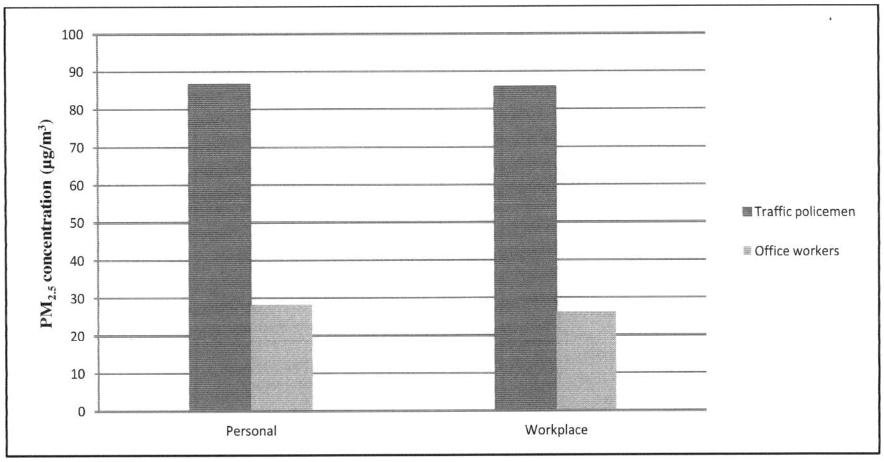

Figure 1: $PM_{2.5}$ exposure among respondents in Klang Valley

In Malaysia environment, the concentrations of air pollutants in the urban area are higher at peak hour during weekdays [12]. According to Jalaludin et al. (2014) concentration of $PM_{2.5}$ in the ambient was higher at morning and afternoon caused by traffic congestion of school bus in the school area. In the meantime, the nature of working hour for traffic policemen is also during peak hour [13]. As a result, this condition causes traffic policemen to be exposed to high level of air pollutants. A previous study by Muhammad et al, (2014) in Malaysia found that traffic policemen experience four times higher concentration of PM compare to the comparative group [14]. However the present study the traffic policemen were exposed to three times higher of $PM_{2.5}$ compared to office workers. According to Shamshad et al. (2015), exposure to air pollutants from the urban area could increase the oxidative stress that can risk respiratory health [15]. Furthermore, study finding by Li et al. (2014) found that exposure to high level traffic-related $PM_{2.5}$ has a positive relationship to DNA damage as an indication early stage of cancer [16]. Hence, the urban workers who working outdoor are expected exposure to high level of TRAP and increase the prevalence of respiratory health symptoms and health risk like cancer.

Table 1. Comparison of PM2.5 Exposure among Traffic Policemen and Office Workers

Variables	Group	Median (µg/m³)	IQR (µg/m³)	Z	p
$PM_{2.5}$ (Personal)	Exposed Comparative	89.00 30.00	41.00 9.00	-15.685	>0.001*
$PM_{2.5}$ (Workplace)	Roadside Office	84.95 24.30	35.00 11.30	-14.519	>0.001*

*Significant at p<0.05, N =195; exposed = 105; comparative = 90

3.2. Comparison of respiratory health symptoms among respondents

The comparison of respiratory health symptoms between traffic policemen and office workers is presented in Table 2. There were no significant difference for phlegm ($\chi 2$ = 0.047, p = 0.828), wheezing ($\chi 2$ = 0.113, p = 0.737) and chest tightness ($\chi 2$ =0.649, p = 0.421) symptoms between both groups. Meanwhile, there is a significant difference in cough symptom ($\chi 2$ = 5.465, p = 0.018) between traffic policemen and office workers. This study finding was different from the previous study done by Muhammad et al. (2014) [14]. In their study, only phlegm shows a significant difference between the exposed group and comparative group ($\chi 2$ = 8.711, p = 0.003). The prevalence ratio (PR) of traffic policemen that had a cough is two times more compared to office workers. A TRAP exposure not only affects an adult's respiratory health but also give a negative impact on children's respiratory health. Sofiah and Jalaludin (2013) reported that children that livings near the busy road, who are expected exposure to high level of $PM_{2.5}$ were had lower lung function compared to those who live far from the busy road [17].

Table 2. Comparison of respiratory health symptoms among respondents

Variables	Exposed (160)	Comparative (149)	χ^2	p-value	PR	95% CI
	Total (%)					
Cough						
Yes	61(38.1)	38(25.5)	5.465	**0.018***	1.800	**1.106 – 2.930****
No	99(61.9)	111(74.5)				
Phlegm						
Yes	36(22.5)	32(21.5)	0.047	0.828	1.061	0.619 – 1.820
No	124(77.5)	117(78.5)				
Wheezing						
Yes	18(11.3)	15(10.1)	0.113	0.737	1.132	0.549 – 2.337
No	142(88.7)	134(89.9)				
Chest tightness						
Yes	22(13.8)	16(10.7)	0.649	0.421	1.325	0.667 – 2.633
No	138(86.2)	133(89.3)				

N=309, *significant at $p<0.05$, significant PR at 95% CI > 1

4. Conclusion

This study suggested that exposure to high level of $PM_{2.5}$ from TRAP would increase the prevalence of a cough among urban workers in Klang Valley. The traffic policemen were exposed to $PM_{2.5}$ at three times higher than the comparative group. In addition, the prevalence of cough episode among traffic policemen was two times higher compared to office workers

References

[1] Wong I C K, Ng Y K, and Lui V W Y 2014 Cancers of the lung, head and neck on the rise: Perspectives on the genotoxicity of air pollution *Chinese Journal of Cancer* **33** 476–480

[2] Roba, Carmen, Zoltán T, Melinda K, Cristina R, and Alexandru O, 2014 Determination of volatile organic compounds and particulate matter levels in an urban area from Romania. *Environmental Engineering and Management Journal* **13** 2261–68.

[3] Jamhari A A, Sahani M, Latif M T, Chan K M, Seng H, Khan M F, & Mohd Tahir N 2014 Concentration and source identification of polycyclic aromatic hydrocarbons (PAHs) in PM_{10} of urban , industrial and semi-urban areas in Malaysia *Atmospheric Environment* **86** 16–27

[4] Fanizza, Carla, Barbara D B, Federica I, Maria E S, Roberto S, Marco I, Marcello F, and Federica I 2018 Analysis of major pollutants and physico-chemical characteristics of $PM_{2.5}$ at an urban site in Rome *Science of the Total Environment* **616** 1457–1468.

[5] Jalava P I, Wang Q, Kuuspalo K, Ruusunen J, Hao L, Fang D, Hirvonen M R 2015 Day and night variation in chemical composition and toxicological responses of size segregated urban air PM samples in a high air pollution situation *Atmospheric Environment* **120** 427–437.

[6] Amil N, Latif M T, Khan M F, and Mohamad M 2016. Seasonal variability of $PM_{2.5}$ composition and sources in the Klang Valley urban-industrial environment *Atmospheric Chemistry and Physics* **16** 5357–5381.

[7] Silva C, Marzari J, Aparecida J, Rocha V, Maria V, and Vargas F 2015 Characterization of an area of reference for inhalable particulate matter ($PM_{2.5}$) associated with genetic biomonitoring in children *Mutation Research/Genetic Toxicology and Environmental*

Mutagenesis **778** 44–55.
[8] Patton A P, Perkinsa J, Zamoreb W, Levyc J I, Brugged D, and Duranta J L 2014 Spatial and temporal differences in traffic-related air pollution in three urban neighborhoods near an interstate highway *Atmos Environ* **99** 309–321
[9] Abidin, E. Z., Semple, S., & Rasdi, I. 2014. The relationship between air pollution and asthma in Malaysian schoolchildren. Air Quality, Atmosphere and Health, 7(4): 421–432.
[10] Bowatte G, Lodge C J, Knibbs L D, Erbas B, Perret J L, Jalaludin B, Dharmage S C 2018 Traffic related air pollution and development and persistence of asthma and low lung function *Environment International* **113** 170–176.
[11] Patil, Rajan R, Satish K C, and Mapilliraju B 2014 Global review of studies on traffic police with special focus on environmental health effects *International Journal of Occupational Medicine and Environmental Health* **27** 523–35.
[12] Latif M T, Dominick D, Ahamad F, Khan M F, Juneng L, Hamzah F M, and Nadzir M S M. 2014 Long term assessment of air quality from a background station on the Malaysian Peninsular *Science of the Total Environment* **482** 336–348.
[13] Jalaludin J, Noh S N S, Suhaimi N F, and Akim A M 2014 Tumor necrosis factor-alpha as biomarkers of exposure to indoor pollutants among primary school children in Klang Valley *American Journal of Applied Sciences* **11** 1616–1630
[14] Muhammad S, Jalaludin J, and Sharmadevan S 2014 Exposure to respirable dust (PM_{10}) and respiratory health among traffic policemen in Selangor *Advances in Environmental Biology* **15** 199–206.
[15] Shamshad S, Indira A P, Shameela S, Chakrapani I S, and Mohammad O A 2015 Oxidative stress in traffic police of Kurnool town *European Journal Of Pharmaceutical and Medical Research* **7** 353–57
[16] Li P, Jinzhuo Z, Changyi G, Liang B, Yuquan X, Haidong K, and Weimin S 2014 Association between individual $PM_{2.5}$ Exposure and DNA damage in traffic policemen *Journal of Occupational and Environmental Medicine* **56** 98–101.
[17] Sofiah N A, and Jalaludin J 2013 Indoor particulate matter 2.5 ($PM_{2.5}$) and lung function among children living near busy road in Cheras Kuala Lumpur *Health and the Environment Journal* **4** 1–19

Acknowledgment
The authors would like to thank the Ministry of Higher Education, Malaysia for supporting the study with a grant under Fundamental Research Grant Scheme (Project Code: 04-01-16-1808FR). We also wish to extend our gratitude to The Royal Malaysia Police of Bukit Aman for the permission and cooperation of conducting this research among traffic policemen in Klang Valley. The authors also declare that there is no conflict of interests.

IOP Conf. Series: Earth and Environmental Science **228** (2019) 012017 doi:10.1088/1755-1315/228/1/012017

Traffic-related Air Pollution (TRAP), Air Quality Perception and Respiratory Health Symptoms of Active Commuters in a University Outdoor Environment

M F Zakaria[1], E Ezani[1]*, N Hassan[2], N A Ramli[1], and M I A Wahab[3]

[1]Department of Environmental and Occupational Health, Faculty of Medicine and Health Sciences, Universiti Putra Malaysia, Serdang 43400 Selangor, Malaysia
[2]Department of Human and Societal Wellbeing, Faculty of Social Sciences and Humanities, Universiti Kebangsaan Malaysia, Bangi 43600 Selangor, Malaysia
[3]Department of Diagnostic and Applied Health Sciences, Faculty of Health Sciences, Jalan Raja Muda Abdul Aziz 50300 Kuala Lumpur, Malaysia

*Corresponding author's email: elianiezani@upm.edu.my

Abstract. Air quality plays significant role in human's health and wellbeing. Vehicle exhaust emissions may lead to health risks especially people who are actively commute. This study aims to examine the association between traffic-related air pollution, perception on traffic pollution and respiratory health symptoms among pedestrian and cyclists in a university campus located in Selangor, Malaysia. A self-administered questionnaire was used to collect data on sociodemographic, air quality perception and respiratory health symptoms among university students (N=180). Air pollutants (PM_{10}, $PM_{2.5}$ and ozone(O_3)) were measured during rush hour (morning, afternoon and evening) simultaneously with traffic count nearby campus roadsides. The average levels of PM_{10} (83.8 μg/m^3), $PM_{2.5}$ (48.9 μg/m^3) and O_3 (314.9 μg/m^3) were higher during rush hour measurements. 51.1% participants agreed that high number of old and private vehicles were the major contributor of air pollution. There were significant associations between each level of traffic-related air pollutants and air quality perception and respiratory health symptoms ($p<0.05$). This study suggests preventive measures for the university management to control traffic-related air pollution in the campus areas.

1. Introduction

Air pollution has recently received greater attention due to its detrimental effects to human health. Malaysia has recorded about 26.3 million vehicles registered by end of year 2015 compared to 20.1 million in 2010, an increase of almost 6.2 million vehicles [1]. This traffic congestion may increase vehicle emissions that contribute to outdoor air pollution especially in urban areas [2]. Traffic-related air pollution is a major contributor to outdoor air pollution in urban settings, producing significant amounts of carbon dioxide (CO2), particulate matter (PM), ozone (O_3), nitrogen dioxide (NO_2), carbon monoxide (CO), polycyclic aromatic hydrocarbons (PAHs), and volatile organic compounds (VOCs) [2].

Outdoor air pollution has adverse health effects on active commuters, especially pedestrian and cyclists worldwide. Nonetheless, study into this phenomenon in Malaysia is rather limited. A study found out that Port Klang, Malaysia suffered from highest PM_{10} concentration as there was heavy traffic congestion due to active industrial activities. People who were living and working in that area were

Content from this work may be used under the terms of the Creative Commons Attribution 3.0 licence. Any further distribution of this work must maintain attribution to the author(s) and the title of the work, journal citation and DOI.
Published under licence by IOP Publishing Ltd

exposed to high traffic pollution levels [3]. Another study observed from monitoring stations in the Klang Valley, Malaysia had found higher levels of PM_{10}, CO, NO_2 and SO_2 with the effects of traffic congestion [4]. Biomass burning activities and ultra violet radiation from sunlight also contributed significantly to high concentrations of PM_{10} and ozone particularly from the tropics region. United States Environmental Protection Act (USEPA) classifies particle pollution mainly into two different sizes based on their predicted penetration capacity into the lung; coarse particulate matter (PM_{10}) and fine particulate matter ($PM_{2.5}$). Coarse particulate matter is a particle with an aerodynamic diameter of 10 μm whereas fine particulate matter with an aerodynamic diameter of 2.5 μm [5]. This type of particulate airborne will be deposited into head region of the airway when inhaled due to its bigger size compared to fine particulate matter which will be deposited further into the tracheobronchial region [6]. The pollutant is generally emitted from vehicle components such as brakes and tires as well as suspension of road dust in urban areas [7].

Active transport could be generalized as any forms of movement that was based on human energy including walking, cycling, scootering, skateboarding, rollerblading and also the use public transport [8]. Promoting these activities in the community may increase their physical activity, decrease pollution level and the impacts, and change the number of accidents. Active travel may cause people having the highest exposure to air pollution as they are directly inhaled the pollutants. A study in the United States had found active travelers were moderately exposed to particulate concentrations especially on major roads whereby pedestrian mostly used passages around activity center such as along main transportation roads which caused them to have high exposure to the pollutants [9]. However, no such studies yet on active travelers' exposure being conducted in Malaysia. In addition, study on short-term exposure to diesel emissions from transport emissions may cause hypersensitive allergies and asthma-like symptoms [10]. The anthropogenic sources primarily come from the automobile exhaust the ambient air pollutants comprise of carcinogenic and mutagenic compounds and lead to possible health risk among the population especially individuals who are active in outdoor. Clearly, outdoor air pollution threatens human health and wellbeing in many ways. Our health can deteriorate when the air quality is poor. Thus, preventive measures are urgently needed in order to overcome the health effects of air pollution.

Human perception plays a fundamental role on people's response to preventive measures [11]. People's behaviors and responses to preventive measures depends on the way they perceive environmental stimuli, therefore to protect public health through adaptation measures it is significant to consider people's perception and behavioral changes. The sense of air quality varies among groups and individuals [12]. Our study examines traffic-related air quality perception among active commuters in an outdoor university campus environment. This study on perception suggests improvement of transportation system in the campus and reducing personal's risk to adverse health effects.

The aim of this study is to examine traffic-related air pollution, perception on air quality and respiratory health symptoms among university students who were actively walking and cycling within campus in one university in Malaysia. The study included measurements of PM_{10}, $PM_{2.5}$ and ozone in different traffic rush hour (i.e. morning, afternoon and evening) at three selected roadsides used frequently by the students. Our study also included measurements on perception air quality and respiratory health symptoms. To the best of our knowledge, this study is the first attempt to examine traffic-related air quality perception among students in a university campus environment in Malaysia.

2. Methodology

Measurements of traffic-related air pollutants (PM_{10}, $PM_{2.5}$ and ozone) were conducted between February and March 2018 at three sampling sites in South Campus in a University (Figure 1). For the same period that the sampling was conducted, a self-administered questionnaire used to collect data on sociodemographic, air quality perception and respiratory health symptoms. The questionnaire was distributed to first-and second-year undergraduate students who were actively cycling and walking from their residential college to attend the class at the academic zones in North Campus. Additionally, traffic-related air pollution measurements were taken in the morning, afternoon and evening alongside traffic count survey to examine the temporal relationship between period of rush hour and traffic volume.

2.1. Study location

The South Campus is part of institutional area sited at southern of University main campus in Serdang, Selangor. This campus is located approximately 28 km southeast from central Kuala Lumpur within the region of Klang Valley and District of Hulu Langat. The roads within the campus can be very crowded with local traffics of daily vehicular movements during peak hours due to its location nearby the main road which connected to academic zones at North campus, student's residential colleges and student health clinic.

2.2. Air sampling

Static ambient measurements for PM_{10}, $PM_{2.5}$ and ozone were carried out using Dustrak Aerosol Monitor 8532 (TSI Inc, Shoreview, MN), Sidepak Personal Monitor AM520 (TSI Inc, Shoreview,MN) and Aeroqual Series 500 (Aeroquál, Auckland, NZ) respectively. These real-time monitors were placed on the bench at the height of ~1.1 m elevation above the ground at each sampling point (JA1, JA2 and JU). All measurements were logged to record 5-min average concentrations at three different rush hour morning (7.00 am -8.00 am), afternoon (1.00 pm -2.00 pm) and evening (5.00 pm – 6.00 pm). To record day-to-day and rush hour differences in background concentrations, the monitoring at a reference site was conducted for 30 minutes before and after rush hour measurements.

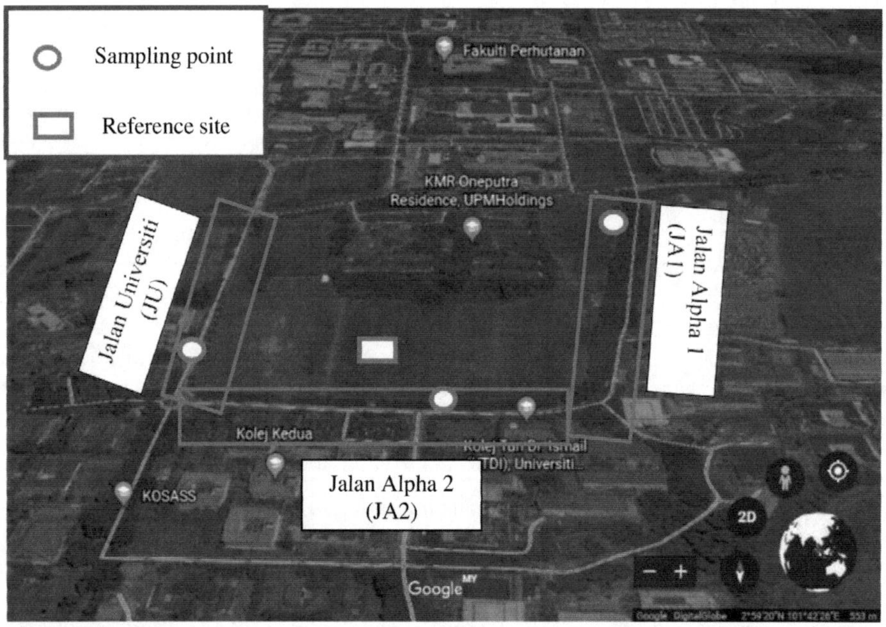

Figure 1. Sampling locations and reference site at Jalan Alpha 1(JA1), Jalan Alpha 2 (JA2) and Jalan Universiti (JU) at South Campus. (Source: Google Earth).

2.3. Subject recruitment

A total of 180 participants were involved in an online survey study. The sample size was calculated using the 2 proportions formula giving a minimum sample size 180 after considering the design effect and 20% of non-response rate [13]. Simple random sampling was used. Participants were invited to participate in the study via social media, including WhatsApp and Facebook. Participants were selected based on specific inclusion criteria; (1) actively cycling or walking within campus; (2) age between 19 and 22; (3) first and second year of study; (4) and living in residential colleges within the sampling points (JA1, JA2 and JU). Those with previous history of respiratory illnesses were excluded.

2.3.1. Questionnaire. The questionnaire was adapted and modified from several studies [14, 15, 16]. It consisted of four sections; i) background characteristics of the respondents (7 items), covering questions on sociodemographic, residential information and environmental tobacco smoke; ii) transportation mode (3 items); iii) air quality perception (3 sub-section with 9 items) and iv) respiratory health symptoms among respondents (6 items). The questionnaire was prepared in Malay and English language and underwent a back-to-back translation to ensure that it suggests similar meaning.

2.4. *Traffic survey*
To assess contribution of main sources of traffic pollutants, a survey of traffic volume by vehicle type was conducted simultaneously with the measurements of PM_{10}, $PM_{2.5}$ and ozone. The manual classified count method used to quantify the traffic volume [17]. The variables were categorized into five categories; motorcycles, cars and taxis, buses, light good vehicles and also heavy good vehicles.

2.5. *Quality control*
All sampling instruments (Dust Trak, Sidepak and Aeroqual) were calibrated by the manufacturer. The sensors were switched on and leave for 10 minutes to ensure it was in a good condition and eliminate any possible errors of data. A pre-test on the questionnaire was held to ensure its comprehensibility. The reliability analysis of each section of the questions showed Cronbach Alpha of 0.7.

2.6. *Data analysis*
Descriptive analysis was computed for each pollutant concentrations measured in different rush hour using Microsoft Excel 2016. The analysis was including frequencies, percentage, mean, median, interquartile range and standard deviation for sociodemographic, air quality perception and level of traffic-related air pollutants. The data collected from questionnaire were analysed using IBM Statistical Package for the Social Sciences (SPSS), version 22.0. Chi square test was undertaken to measure association between the categorical independent variables and health effects. Level of significance was set at 0.05.

2.7. *Ethical approval*
Ethics approval was obtained from the University Ethic Committee for Research Involving Human Subjects with the reference number of JKEUPM-2017-197. Informed consent was attached with the questionnaire and distributed to all participants. The anonymity of the participants was maintained at all times.

3. Results

3.1. *Traffic-related Air Pollution and Traffic Count*
The full list of three traffic-related air pollutants (PM_{10}, $PM_{2.5}$ and ozone) we quantified at three sampling points shown in Table 1. Measurements at three rush hours (morning, afternoon and evening) were also shown below. The sampling campaign was conducted between February and March 2018. This has resulted in 41 sampling counts with 15 number of measurements were conducted during morning, 14 in the afternoon and 12 in the evening rush hour. Measurements of PM_{10}, $PM_{2.5}$ and ozone made at JA2 and JU were higher than JA1. This is because the location of JA2-JU is located at intersection with high density of traffic and may contribute to high concentrations of particulate matter (PM) compared to JA1.

Concentration levels of PM_{10} (average morning max±SD= 87.7±18.2 $\mu g/m^3$, average afternoon max±SD= 77.8±26.1 $\mu g/m^3$) and $PM_{2.5}$ (average morning max±SD= 44.1±12.7 $\mu g/m^3$, average afternoon max±SD= 49.0±13.1$\mu g/m^3$) were highest during morning and afternoon rush hour. This suggests the airborne particulate pollutants concentrations were coincided with each other and originated from the source. In contrast, ozone levels were found highest during afternoon and evening rush hour.

Both of PM_{10} (average max±SD= 83.8±18.2 μg/m^3) and ozone (average max±SD= 315.0±137.8 μg/m^3) were exceeded standards set by World Health Organisation (WHO) and Malaysian Air Quality Standards (MAQS) respectively.

Table 2 shows the traffic volume in three sampling locations; Jalan Alpha 1 (JA1), Jalan Alpha 2 (JA2) and Jalan Universiti (JU). Total number traffic count was higher at JA2 and JU. This has coincided with higher average concentrations of PM_{10} and $PM_{2.5}$ measured at both roadsides. Motorcycles, car and taxis were found dominant on JA2 and JU. Number of buses was found highest at JA1. This may because JA1 is the main entrance to the South campus, therefore it had been the main road that connect the academic zone to all residential colleges located at this campus area.

Table 1. Mean and standard deviation of PM_{10}, $PM_{2.5}$ and ozone for three periods of rush hour measured at roadsides within South Campus during 1-h measurement.

Location	Rush hour	PM_{10}	$PM_{2.5}$	Ozone
		Mean (SD) (μg/m^3)		
Jalan Alpha 1 (JA1)	Morning	72.4 (30.3)	34.1 (9.7)	22.8 (8.1)
	Afternoon	58.3 (36.3)	25.7 (21.1)	194.3 (171.7)
	Evening	39.3 (36.5)	21.6 (21.6)	159.0 (165.6)
Jalan Alpha 2 (JA2)	Morning	83.8 (18.2)	49.0 (13.1)	31.4 (8.8)
	Afternoon	70.6 (35.6)	35.9 (17.1)	303.0 (234.1)
	Evening	53.0 (36.9)	26.8 (17.4)	315.0 (137.8)
Jalan Universiti (JU)	Morning	76.5 (26.9)	40.7 (14.12)	21.8 (4.9)
	Afternoon	77.8 (26.1)	44.1 (12.7)	127.5 (78.3)
	Evening	38.8 (27.9)	43.5 (72.3)	180.5 (147.4)

Table 2. Traffic count of different type of vehicles within sampling sites located at South Campus area between February and March 2018.

Sampling site	Observational traffic volume					
	Total	Motorcycles	Cars and taxis	Buses	[a]LGV	[b]HGV
Jalan Alpha 1 (JA1)	2299	881	1286	91	41	0
Jalan Alpha 2 (JA2)	2760	1190	1526	2	39	3
Jalan Universiti (JU)	2886	1338	1507	3	37	1

[a]Light Good Vehicles [b]Heavy Good Vehicles

3.2. Surveys on traffic-related air pollution perception

3.2.1. Socio demographic. A total of 184 questionnaires being answered, 4 were excluded due to incomplete questionnaire returned, giving the response rate 100%. Table 3 shows characteristics of participants by first-and second year undergraduates during the questionnaire surveys. Majority of participants were 21 to 22 years old, Malay ethnic, female students and second-hand smokers. There was equal number of respondents who were cycling and walking from their residential colleges at South Campus to academic zones at North Campus using JA1, JA2 and JU.

Table 3. Participants' demographic information.

Characteristics	Pedestrian (n=90)	Cyclist (n=90)
	n (%)	
Age (years old)		
19 - 20	43	44
21 - 22	46	46
> 22	1	0
Gender		
Male	29	43
Female	61	47
Ethnicity		
Malay	72	45
Chinese	9	25
Indian	5	15
Others	4	5
Year of study		
First year	42	44
Second year	48	46
Smoking		
Yes	5	3
No	85	87
Second-hand smokers		
Yes	13	12
No	77	78

3.2.2. Perception on traffic pollution. The survey covers; student's perception on the air quality in the campus area, sources of traffic pollution and the strategies to mitigate air pollution. Among the 180 respondents, 89 (49.4%) of respondents believed that air quality of university campus area was better compared to six months ago, while 48 (26.7%) believed that air quality was worse since six months ago. The air quality rating is presented in Figure 2.

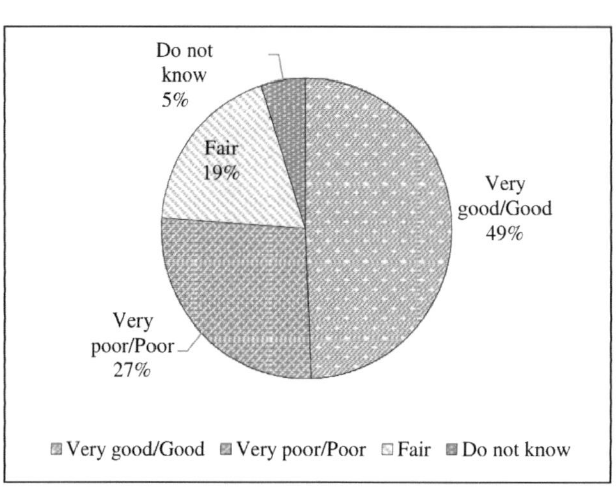

Figure 2. Overall air quality rating in campus area.

As shown in Figure 3, majority (83.4%) of the participants agreed that the problem of traffic pollution was ‘old vehicles’ in campus area. They also agreed that ‘too many private vehicles’ (51.1%) also contributed to traffic pollution. Moreover, participants did not agree that (68.9%) ‘traffic congestion’ and (81.1%) ‘diesel vehicles’ automatically influenced the vehicles emission significantly impacted the level of traffic pollution.

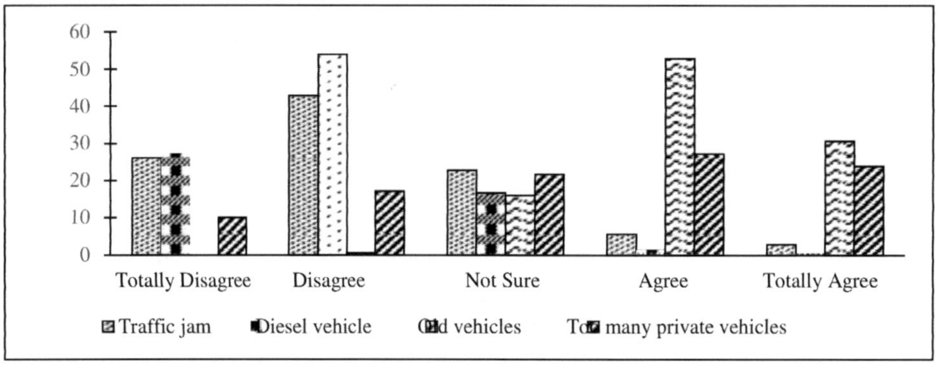

Figure 3. Aspects of motor vehicles affect air pollution.

Participants were also asked of their perceptions on how to mitigate traffic pollution, Figure 4 shows that highest percentage of respondents (73.3%) agreed that by improving of pedestrian and cyclists’ lane facilities would be one of the measures to reduce air pollution. Majority of respondents agreed that carpooling (68.9%), reducing number of private vehicles (66.7%) and also walking and cycling to class (66.1%) will decrease the level of air pollution in the campus area.

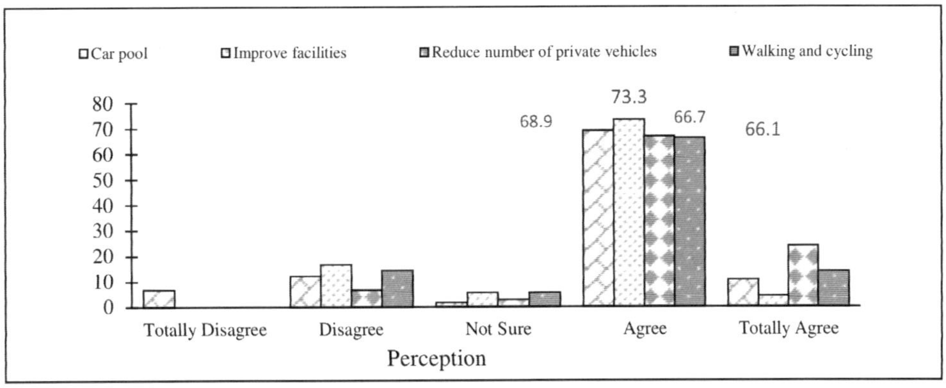

Figure 4. Initiatives to control and reduce air pollution

3.3. Association TRAP with air quality perception and respiratory health symptoms

To associate traffic-related air pollutions with air quality perception, all pollutants measured were categorized based on their median value and divided by low and high. This test only involved those who perceived air quality as poor (n=60) and good (n=77). Table 4 shows the association between traffic-related air pollutions and air quality perception among students who were cycling and walking within study locations. The Pearson's chi square test results show a significant association between all pollutants (PM_{10}, $PM_{2.5}$, and ozone) with air quality perception among respondents as (p= 0.001, 95% CI= 3.877-46.347), (p= 0.001, 95% CI= 3.877-46.347) and (p= 0.011, 95% CI= 0.188-0.811) respectively.

Table 4. Association between traffic-related air pollutions with air quality perception among respondents using Pearson's chi-square test (N=137).

Air Pollutant	Air Quality Perception: Poor Air Quality (n=48)	Good Air Quality (n=89)	χ^2 value	p value	95% CI
$PM_{2.5}$					
High	45	47	23.196	**0.001***	3.877-46.347
Low	3	42			
PM_{10}					
High	45	47	23.196	**0.001***	3.877-46.347
Low	3	42			
Ozone					
High	24	64	6.516	**0.011***	0.188-0.811
Low	24	25			

Further analysis was done to determine an association between traffic-related air pollutions exposure with respiratory health symptoms among the active travellers by using the median value each measured air pollutant. The results were shown in Table 5. There is a significant association between high level of $PM_{2.5}$ and PM_{10} with five respiratory health symptoms which are wheezing (p= 0.000, 95% CI= 0.021-

0.107), chest tightness (p= 0.000, 95% CI= 0.052-0.225), shortness of breath (p=0.000, 95% CI= 0.060-0.340), coughing (p= 0.000, PR= 0.158, 95% CI= 0.079-0.315) and phlegm (p=0.006, 95% CI= 0.0163-0.748). A significant association between was also observed with high level of ozone and five respiratory health symptoms; wheezing (p= 0.002, 95% CI= 1.464-5.541), chest tightness (p= 0.000, 95% CI=2.141-8.354), shortness of breath (p= 0.001, 95% CI= 1.557-6.283), phlegm (p= 0.041, 95% CI=1.029-4.299) and breathing difficulties (p= 0.039, 95% CI= 0.252-0.972).

Table 5. The association between exposure of (a) $PM_{2.5}$, (b) PM_{10} and (c) ozone with respiratory health symptoms among respondents using Pearson's chi-square test (N=180).

(a) Respiratory Symptoms	$PM_{2.5}$ (High)	$PM_{2.5}$ (Low)	χ^2 value	p value	95% CI
	Number				
Wheezing					
Yes	74	9	68.297	**0.001***	0.021-0.107
No	27	70			
Chest Tightness					
Yes	63	12	40.607	**0.001***	0.052-0.225
No	38	67			
Shortness of Breath					
Yes	41	7	22.826	**0.001***	0.060-0.340
No	60	72			
Coughing					
Yes	85	36	29.458	**0.001***	0.079-0.315
No	16	43			
Phlegm					
Yes	32	11	7.689	**0.006***	0.163-0.748
No	69	68			
Breathing Difficulties					
Yes	34	21	1.047	0.306	0.733-2.679
No	67	58			

(b) Respiratory Symptoms	PM_{10} (High)	PM_{10} (Low)	χ^2 value	p value	95% CI
	Number				
Wheezing					
Yes	74	9	68.297	**0.001***	0.021-0.107
No	27	70			
Chest Tightness					
Yes	63	12	40.607	**0.001***	0.052-0.225
No	38	67			
Shortness of Breath					

Yes	41	7	22.826	**0.001***	0.060-0.340
No	60	72			
Coughing					
Yes	85	36	29.458	**0.001***	0.079-0.315
No	16	43			
Phlegm					
Yes	32	11	7.689	**0.006***	0.163-0.748
No	69	68			
Breathing Difficulties					
Yes	34	21	1.047	0.306	0.733-2.679
No	67	58			

(c) Respiratory Symptoms	Ozone (High)	Ozone (Low)	χ^2 value	p value	95% CI
	Number				
Wheezing					
Yes	49	34	9.838	**0.002***	1.464-5.541
No	78	19			
Chest Tightness					
Yes	40	35	18.356	**<0.001***	2.141-8.354
No	87	18			
Shortness of Breath					
Yes	25	23	10.751	**<0.001***	1.557-6.283
No	102	30			
Coughing					
Yes	81	40	2.320	0.128	0.848-3.600
No	46	13			
Phlegm					
Yes	25	18	4.912	**0.041***	1.029-4.299
No	102	35			
Breathing Difficulties					
Yes	33	22	4.248	**0.039***	0.252-0.972
No	94	31			

*significant at $p<0.05$

4. Discussion

Traffic related air pollution was measured in three different locations in a university campus. Average concentrations of PM_{10}, $PM_{2.5}$ and ozone were used to differentiate levels of air pollution during rush hour i.e. morning, afternoon and evening. A set of validated questionnaires was also distributed to first- and second year undergraduates who were actively commute to academic zones from his/her residential colleges either cycling or walking. This was to investigate traffic-related air pollution perception and respiratory health effects. Pearson chi-square was applied to explore traffic-related air pollutants levels and association between perception and health condition.

In general, levels of particulate air pollutants (PM_{10} and $PM_{2.5}$) were higher during morning and afternoon and in turn corresponded with high number of traffic count at three sampling points (JA1, JA2 and JU). This suggests ambient sources may drive from exhaust emissions from vehicles including private cars, taxis and buses that passed by the roadsides. PM measurements were found higher at near roadsides in which may influence by density of road networks and distance from emission sources [8]. The spatial patterns between traffic pollutants were similar although the levels of spatial variability were diverges [8]. Ozone levels were recorded highest during afternoon and evening measurements. Ozone measurements were the lowest in the morning. This corresponds to the solar intensity and oxidation process of precursor pollutants emitted into the atmosphere (NO_x) during early morning hour. The levels of ozone will accumulate and increasing during midday and evening peak.

Our study explored traffic-related air pollution on university student's perception. Majority of the participants (83.4%) agreed that air pollution was motor vehicles. This is in line with study conducted in Kuala Lumpur city centre in which the main source of air pollution was from traffic emissions [18]. Our results also agreed with a study on air quality perception in the United Kingdom, in which majority of their respondents perceived automobiles as the main source of air pollution [19]. In contrast, a study on air quality perception Kuwait found that people identified air pollution to be the fifth most important consequence of traffic congestion out of eight possible consequences [20]. Majority of our respondents agreed to suggest that car pool, improving facilities of pedestrian and cyclists' lane, reduced the number of private vehicles and also walking and cycling to the academic zones as control measures to reduce air pollution in campus area. This suggesting the role of practical everyday experience that may influence on how people perceived the quality of air especially to the active commuters.

Most studies have highlighted that the real-world experience of the respondents including the smells of pollutants and what they have seen were the major influence for their perception on the air pollution levels [19, 21]. Similar study also conducted whereby all respondents were randomly chose to answer their perception and awareness on traffic pollution [18]. However, not much work has been done to study how human perception correlates with physical measurements of pollution and their experience as cyclist or pedestrian. Our study has found a significant association between air quality perception with air pollutants (PM_{10}, $PM_{2.5}$ and ozone) ($p<0.05$). This has in line with observations made by researchers in China which found the levels of ozone and visibility were more correlated with perceived air quality [14]. People who were actively commuted are easily exposed to traffic emissions and might harm their health outcome. Respiratory diseases including asthma may affect pedestrian and cyclist during their active travel. Our study found levels of PM_{10}, $PM_{2.5}$ and ozone measured have significant associated with respiratory symptoms (wheezing, shortness of breath, coughing and phlegm) among students.

We have identified several study limitations. Our study was conducted through cross-sectional design, therefore we cannot predict whether certain variables precedes the other. It is also noted that our study used limited number of sample size; air monitoring exercise and participants. Our study adopted self-reported on respiratory symptoms, therefore recall bias might arise among respondent. Non-invasive health measurement such as lung function test should be included in the future studies. Due to short duration of the project we were unable to perform the assessment and validation of three portable real-time monitors (i.e Dustrak, Sidepak and Aeroqual S500) prior to our sampling campaign. However, this type of sampling exercise is important to examine outdoor exposure assessment in a representative campus area environment. Additional meteorological measurements such as relative humidity, temperature and wind speed should take into account for future research.

5. Conclusion

Our findings have revealed a significant association between air pollution concentrations and air quality perception among students who actively commute within the university campus area. It reveals that respiratory health symptoms are affected by exposure to traffic related air pollution. Most of participants strongly agreed to policies and strategies in reducing air pollution in the university campus. For instances, reducing the number of private vehicles, carpooling, improving facilities for pedestrian and cyclists and sustaining healthy lifestyle in the campus. Therefore, these strategies should be taken into consideration by the university management in formulating policies, measures to mitigate outdoor air pollution by reducing traffic-related air pollution exposure to active commuters within the campus area.

References

[1] Road Transport Department Malaysia 2017 Transport Statistic Malaysia p 6
[2] Health Effects Institute 2010 Traffic-Related Air Pollution: A Critical Review of the Literature on Emissions, Exposure, and Health Effects *Special Report* **17** p 2
[3] Mohamad ND, Ash'aari ZH and Othman M 2015 Preliminary Assessment of Air Pollutant Sources Identification at Selected Monitoring Stations in Klang Valley, Malaysia *Procedia Environmental Sciences* **30** p 121–126
[4] Azmi ZF, Latif MT, Ismail AS, Juneng L and Jemain AZ 2010 Trend and status of air quality at three different monitoring stations in the Klang Valley, Malaysia *Air Quality, Atmosphere and Health* **3(1)** p 53-64
[5] United States Environmental Protection (USEPA) 2016 [accessed online:https://www.epa.gov/pm-pollution/particulate-matter-pm-basics#PM]
[6] Brown JS, Gordon T, Price O and Asgharian B 2013 Thoracic and respirable particle definitions for human health risk assessment. *Particle and Fibre Toxicology* **10** p 1-12
[7] Kim H, Kabir E, and Kabir S 2015 A review on the human health impact of airborne particulate matter *Environment International* **74** p 136–143
[8] Lorimer SW and Marshall S 2015 Beyond walking and cycling: scoping small-wheel modes *Proceedings of the Institution of Civil Engineers* p 1-10
[9] Hankey S and Marshall JD 2017 Urban Form, Air Pollution, and Health *Curr Environ Health Rep*. **4(4)** p 491-503
[10] Donaldson K, Gilmour MI and MacNee W 2000 Asthma and PM(10) *Respiratory Research* **1(1)** p 12-5
[11] Pantavou K, Lykoudis S and Psiloglou B 2017 Air quality perception of pedestrians in an urban outdoor Mediterranean environment: A field survey approach *Science of the Total Environment* **574** p 663–670
[12] Van Kamp I, Leidelmeijer K and Marsman G 2003 Urban environmental quality and human well-being towards a conceptual framework and demarcation of concepts vol 65 pp 5–18
[13] Lwanga S and Lemeshow S 1991 Sample size determination in health studies: a practical manual (England: World Health Organization)
[14] Liao X, Tu H, Maddock JE, Fan S, Lan G, Wu Y, Yuan ZK, Lu Y 2015 Residents' perception of air quality, pollution sources, and air pollution control in Nanchang, China *Atmospheric Pollution Research* **6** p 835-41
[15] Deguen S, Ségala C, Pédrono G, Mesbah M 2012 A new air quality perception scale for global assessment of air pollution health effects. Risk Analysis: An International Journal 32 pp 2043-2054
[16] International Union against Tuberculosis and Lung Diseases (IUALTD) 2001 The international journal of tuberculosis and lung disease *The official journal of the International Union against Tuberculosis and Lung Disease* **5** p 213-215
[17] Zheng P and Mike M 2012 An investigation on the manual traffic count accuracy *Procedia-Social and Behavioral Sciences* **43** p 226-231
[18] Shafie SH and Mahmud M 2017 Public perception on traffic pollution in federal territory of Kuala Lumpur, Malaysia Planning Malaysia Journal 15 pp 105-114
[19] Howel D, Moffatt S, Bush J, Dunn CE, Prince H 2003 Public views on the links between air pollution and health in Northeast England *Environmental Research* **91** p 163-71

[20] Koushki PA, Ali MA, Chandrasekhar BP, Al-Sarawi M 2002 Exposure to noise inside transit buses in Kuwait: Measurements and passenger attitudes *Transport Reviews* **22** p 295-308
[21] Bickerstaff K and Walker G 2001 Public understandings of air pollution: the 'localisation'of environmental risk *Global Environmental Change* **11** p 133-145

Acknowledgement
The authors would like to thank the Department of Environmental and Occupational Health, Faculty of Medicine and Health Sciences, Universiti Putra Malaysia and Faculty of Health Sciences, Universiti Kebangsaan Malaysia for equipment renting.

What Makes a Location into a "Favorable Habitat" under Changing Climate and Environmental Conditions? A Pilot Study Focused on Exploring the Differences between Natural and Non-natural Habitats using Airborne LiDAR

M Mohan [1,2,*], G P Catts [1], B Vaughan [3], J P Roise [1,2], C A Silva [4,5], J B McCarter [1], P. Jat [6], W S Wan Mohd Jaafar [7,8], G Gopan [9], K N Abdul Maulud [8,10]

[1]Department of Forestry and Environmental Resources, North Carolina State University, 2800 Faucette Drive, Raleigh, NC 27695, USA
[2]Department of Operations Research, North Carolina State University, 2310 Stinson Drive, Raleigh, NC 27695, USA
[3]Forest Financial Planning and Analysis, The Conservation Fund, Chapel Hill, NC 27514, USA
[4]NASA Goddard Space Flight Center, 8800 Greenbelt Road, Greenbelt, MD 20771, USA
[5]Department of Geographical Sciences, University of Maryland, College Park, Maryland, MD 20740, USA
[6]Department of Environmental Sciences and Engineering, UNC-Chapel Hill, 135 Dauer Drive, Chapel Hill, NC 27599, USA
[7]School of Geosciences, University of Edinburgh, Edinburgh EH8 9XL, UK
[8]Earth Observation Centre, Institute of Climate Change (IPI), Universiti Kebangsaan Malaysia, Bangi, Selangor, Malaysia.
[9]Department of Earth Sciences, National Institute of Technology, G.E. Road, Raipur, Chhattisgarh 492001, India.
[10]Smart and Sustainable Township Research Centre (SUTRA), Universiti Kebangsaan Malaysia, Bangi, Selangor, Malaysia.

Email: mmohan2@ncsu.edu

Abstract. Recovery plans supported by conservation models are critical for the protection of endangered species. For developing these models, parameters are most commonly extracted via field survey or remote sensing based methods. However, at times, these models get narrowed down to specific habitat features associated with naturally occurring ecosystems, and thereby fail to detect suitable non-natural habitats that the animals have gotten adapted to in recent years - as a survival mechanism to cope up with the dynamic climate and environmental conditions. As a first step to address this issue, we considered the case of Red-cockaded woodpecker (RCW) species and undertook a pilot study to

Content from this work may be used under the terms of the Creative Commons Attribution 3.0 licence. Any further distribution of this work must maintain attribution to the author(s) and the title of the work, journal citation and DOI.
Published under licence by IOP Publishing Ltd

IOP Conf. Series: Earth and Environmental Science **228** (2019) 012018 doi:10.1088/1755-1315/228/1/012018

explore the characteristics of non-natural locations that make it favorable for RCW nesting. On exploring the differences in habitat characteristics of natural and non-natural ecosystems - by employing airborne laser scanning data and logistic regression analysis method - we identified new prominent forest attributes and their variations for each habitat types. Based on our findings, we provide fruitful interpretations, recommendations and encourage discussions on less-studied conservation aspects and hope to stretch the horizons of ongoing biodiversity conservation efforts in the wake of global environmental change.

1. Introduction

Climate change, which broadly refers to the continuous change in climate patterns - evident at a local as well as global scale - primarily due to the increased levels of carbon dioxide in the atmosphere, has been transforming the environment at a dynamic rate, thereby, intervening in the life cycles of both humans and animals equally [1]. As a consequence, several animals respond to these changing environmental conditions by adapting themselves to the newly formed non-natural ecosystems, while numerous species flee or perish. Therefore, conservation models, which are critical for enhancing recovery operations of animals, need to be periodically modified to accommodate the varying habitat features. Nevertheless, these models, which are mostly built using habitat parameters derived through field surveys and remote sensing techniques, often get narrowed to specific habitat features associated with naturally occurring ecosystems, limiting the efficiency of the recovery processes.

Red-cockaded woodpeckers (RCW) (*Picoides borealis* Vieillot) are a group of non-migratory birds that are endemic to southeastern United States. Couple of decades back, in the Piedmont and coastal plains, they were very common throughout the open longleaf pine (Pinus palustris Mill) ecosystems [2]. However, RCWs' preference for longleaf pine and their habitat destruction from the start of 17th century, caused by European settlement, naval stores, urbanization, agriculture and widespread commercial timber harvesting along with issues associated with change in environmental conditions, that caused increased fire suppression, forced the RCWs to be placed on the federal list of endangered species [3]. In the past few years, several RCW habitat conservation activities have been performed by the US Forest Service and similar agencies in the southeastern U.S [4,5]. Yet, most of these efforts were focused on modeling foraging zones of the "natural habitats", which comprises of long leaf ecosystems with low (< 3 m) understory [6]. In these habitat models, the "non-natural" habitat environments, such as pocosins and pond pine ecosystems, where there have been RCW citing, fail to qualify for conservation initiatives.

Discerning the reasons for species adaptability in newly formed ecosystems can extensively assist in identifying and protecting suitable non-natural habitats, and in that way help enhance the ongoing habitat conservation efforts. However, different organisms behave differently to climate and environmental changes and quantifying the dynamic processes attributed to habitat favorability can be perceived as a challenging and time-consuming endeavor. Nevertheless, somewhere, we have to start, in order to progress. This pilot study is performed as a first step towards addressing this salient issue - using Light Detection and Ranging (LiDAR) data and logistic regression analysis method - and by taking the case of RCWs, as a representative species, we hope to stretch the horizons of conservation perspectives and encourage future research activities in directions that further explore the importance of upgrading habitat conservation models under changing climate and environmental conditions.

2. Materials and Methods

2.1. Study Area, Data Collection and Processing

Two study sites were considered as a part of this research study: Palmetto-Peartree Preserve (P3) and Croatan national forest (CNF), both located at North Carolina, USA. Study plots (i.e., trees with RCW cavities) at P3 comprised of non-natural ecosystems and at CNF, the habitats primarily had pine forests. The P3 was in fact established by The Conservation Fund in 1999 to serve as an RCW mitigation bank to offset habitat loss happening from road construction projects. Active RCW nesting cavity locations that had been spotted within P3 were obtained via The Conservation Fund and U.S. Fish and Wildlife Service. In case of CNP, data were acquired through North Carolina Natural Heritage Program 2016, Division of Land and Water Stewardship. From natural habitats, non-natural habitats, and no-bird zones, 60 locations (20 locations from each category) were randomly selected for building the predictive models. Here, we define no-bird zones as areas that the birds have avoided for nesting, but, are similar to the non-natural habitats in terms of physical geography. These zones were selected from regions which were far (at least 8.05 km; i.e., 5 miles) away from identified RCW habitats – as the average flight distance of these birds were found to be less than 8 km. Also, we decided not to include the map of exact study locations in this paper as a gesture in support of the conservation endeavors.

Airborne LiDAR data used for this study were acquired in 2014 as a part of the NC Floodplain Mapping Program (Department of Public Safety). Field survey checkpoints (around 100 checkpoints per county) were used to evaluate the accuracy of LiDAR products; more details can be found in [7]. The major output derived from the LiDAR point cloud was the canopy height model (CHM); canopy structure metrics were derived from these CHMs. First, the LiDAR point clouds associated with the study sites were extracted and Digital Terrain Models (DTMs) were generated, from the points classified as ground. Subsequently, the point clouds were normalized to height above ground by subtracting the DTM elevation from each return, for generating Digital Surface Models (DSMs). The difference between DSMs and DTMs provided us with the respective CHMs. All LiDAR data processing were performed in ESRI ArcGIS 10.3 [8,9], by employing default settings of general raster processing tools such as LAS to Multipoint, Terrain to Raster and Raster Calculator function.

2.2. Spatial Analysis and Habitat Attributes Estimation

In case of RCW, spatially explicit information on the habitat distribution and structure of forest vegetation is required at broad scales for answering questions regarding their habitat selection criteria [6,7]. As an attempt to better understand the spatial pattern of habitat distribution and to improve the conservation practices, various classification and cluster analysis methods had been evaluated in previous studies [11]. Herein, we performed detailed visual analysis of non-natural study spots with the help of RCW experts, and this prompted us to assess the possibilities of having height difference between tree-crown and mid-story as a crucial factor for habitat suitability. Hence, in our habitat presence models we used height difference (d) as a primary variable, along with three other environmental variables - height density factor (f), percent openness (p) and tree height (h) - for quantifying the change in canopy cover with varying heights; parameters are explained in the following sections.

Initially, we developed forest CHMs for the randomly selected individual study sites for extracting the parameters of interest from the LiDAR data. We treated these individual fragments as a circular piece of land having a 30.34 m (100 feet) radius, so as to maintain uniformity and to standardize our values for making easier and fair comparisons. This kind of a smaller circumscribed circle around cluster center was used instead of the 0.8 km (½-mile) radius circle as adopted in the 2003 recovery plan, as our primary objective was the assessment of nesting sites [4,12]; spatial separation of foraging and non-foraging habitats was not considered in the model. Also, parameters associated with midstory and understory conditions had to be compromised as the accuracy and details contained by point-cloud data were low. Tree height, h, in our case, represented the maximum tree height within the circular plot. Since we were not able to clearly quantify the hardwood midstory as "sparse and less than 2.1m in height" as listed in the recovery standards [12], we derived a novel surrogate factor - "height-density factor (f)".

For quantitatively studying the vertical canopy structure, we first built conditional rasters from CHMs - using con tool provided by spatial analyst toolkit available in ArcGIS 10.3 [8,9] - and calculated the percent canopy cover difference for every 3.05 m (10 feet) interval starting from h to the ground level. The Con tool works in a way such that, the output value for each cell is based on whether the cell value is evaluated as true or false (with respect to a conditional statement that we specify; for example: tree height > 10 m). At first, the interval width was also tested for 0.31 m (1 foot), 0.91 m (3 feet) and 1.52

m (5 feet), nonetheless, the canopy cover change with respect to interval width was comparatively less evident in these cases. Most of the times, plotting canopy height with relative difference in canopy cover over the intervals resulted in break points where a rapid change (maximum variation) was observed. Height-density factor can be defined as the center point of this height interval exhibiting maximum variation. Center point was selected as the variation of percent canopy cover within the 3.05 m (10 feet) intervals was found to be gradual. In short, *f* gives us an estimate on the change in canopy cover with respect to varying canopy height and is the canopy height where the % change in canopy cover is maximum. Whereas, the parameter *p* gives us the percent of unoccupied pixels in the CHM where canopy height is equal to *f* and parameter *d* denotes the absolute difference of *h* and *f*.

2.3. Quantitative Analysis and Habitat Modeling

Taking into account the binary nature of RCW tree cavity data, logistic regression and discriminant analysis were considered as appropriate modeling approaches. Nevertheless, logistic regression was considered as more fitting as available data on some of the independent variables (e.g. canopy openness) are qualitative and non-multivariate normal [13]. However, before that, univariate statistical tests were performed to evaluate the validity of individual model variables - and for obtaining a better understanding of each variable's role. If the species did discriminate among sites based on the considered environmental factors, the mean values of these variables at favorable habitat locations should be different from locations they have avoided i.e., no-bird zones [13].

Logistic regression is a predictive analysis method, which is considered appropriate to conduct when the dependent variable is dichotomous (binary). In our case, the logistic model is used to estimate the probability (P) of RCW occurrence/favorability (binary response) based on *h, f, p* and *d* (independent variables). Unlike ordinary multiple regression, logistic regression analysis is efficient when applied to binary data and yields estimated probabilities - called logits - that lie between 0 and 1 (equation 1, 2 & 3). This way, it allows us to evaluate the influence of an individual factor - using the measure of association, the odds ratio - in increasing the probability (by a specific percentage) of a given outcome. An odds ratio indicates how likely - with respect to odds - a particular event occurs in one group with reference to its occurrence in another group [14]. The general form of the equations are as follows:

$$\text{logit}(P_i) = \beta_0 + \beta_1 X_1 + \beta_2 X_2 + \beta_3 X_3 + \beta_4 X_4 + \varepsilon_i . \quad (1)$$

$$\text{logit}(P_i) = \log\,[P_i/(1 - P_i)] \quad (2)$$

$$P_i = 1/\,[1 + e^{\text{logit}(P_i)}]. \quad (3)$$

Where, P_i is the probability that the event (active RCW habitat, for example) occurs in the ith case, log is the natural log (to the base e), logit (P_i) is the logit transformation of the probability of the event, *i* is the indices of all cases (observations), β_0 is the intercept of the regression line, β_1, β_2, β_3 and β_4 are the slopes of the regression lines, X_1, X_2, X_3 and X_4 are the predictor variables, ε_i is the error (residual) associated with each observation and e is the base of the natural logarithms. In our analysis, the probability cut level of P_i was chosen to be 0.50 and with this threshold, the yielded misclassification errors were found to be very limited.

For understanding the influence of individual parameters towards habitat favorability - along with determining the predominant attributes - we developed two separate habitat models. The first model was constructed on parameter values derived from 20 non-natural nesting sites and 20 no-bird zones located at P3. The second model was similar to the first model, but had additional 20 natural sites from CNF. Figure 1 provides an overview of the workflow. Statistical analysis and data exploration were carried out and tested in a combination of R and SAS environment [15,16].

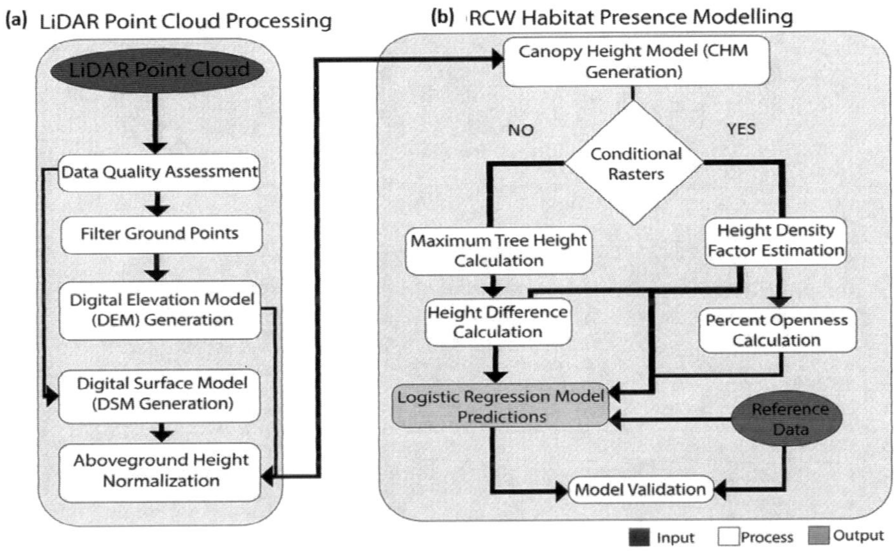

Figure 1. Workflow of LiDAR parametric extraction and habitat modeling

3. Results

3.1. LiDAR derived metrics for describing forest structure

Primary parameters of interest for our study - *h*, *f*, *p*, and *d* - were derived from CHMs. For demonstration purpose, here, we considered the case of a non-natural sample study site (SSS) with the maximum tree height (*h*) being 28.13 m (92.3 feet). For calculating *f*, the CHM was first split into multiple bins of 3.05 m (10 feet) intervals. Here, for the SSS, since the *h* was 28.13 m (92.3 feet), 9 conditional rasters were created - using point clouds greater than 27.43 m (90 feet), 24.38 (80 feet), 21.34 (70 feet) and so on, up to 3.05 m (10 feet). Figure 2 reveals how the canopy cover varies - i.e., gets denser - with decreasing canopy height. Figure 3a shows the percent openness (100 – % canopy cover) - an indicator of habitat favorability - associated with canopy heights. When the relative percent difference in % openness is plotted against canopy height (figure 3b), we notice an abrupt change in pattern when the canopy height is 12.19 m (40 feet). This indicates that the canopy cover was getting denser and denser gradually till 12.19 m (40 feet), and then there occurred a sizeable change. Since, the change in canopy cover is gradual, it can be presumed that the interval (9.14 m to 12.19 m i.e., 30 feet to 40 feet) is where the maximum relative change in canopy cover happened. For the same reason, we choose *f* to be the mid-point of the interval (i.e., 35 feet or 10.67 m); *f* can be perceived as a satisfiable surrogate to the mid-story height. Values of other parameters such as percent openness (*p*) and height difference (*d*), associated to height-density factor estimates, were 68.83 and 17.47 m i.e., 57.3 feet (92.3 – 35) respectively; similar procedures were followed for rest of the study sites as well.

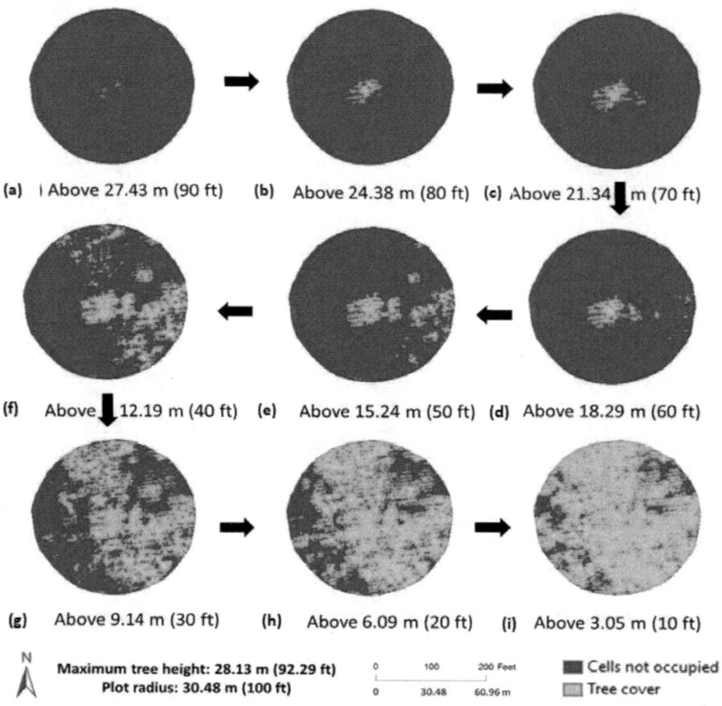

Figure 2. Conditional rasters showing the variation of canopy openness for the sample study site

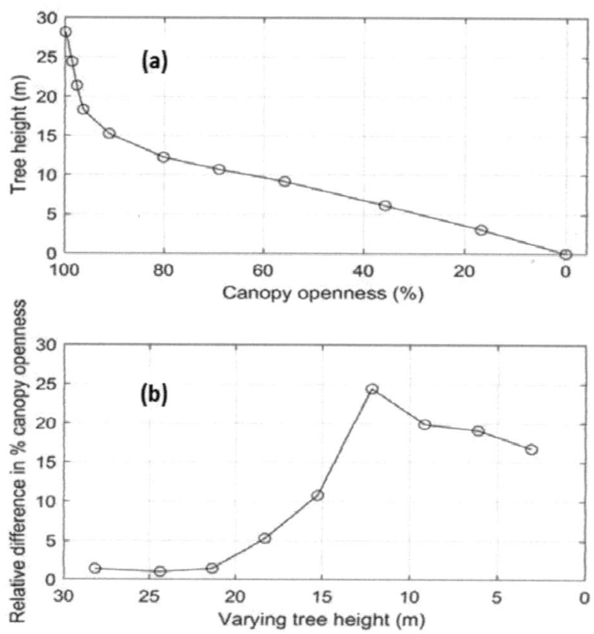

Figure 3. Graphs showing (a) percent canopy openness vs tree height (b) tree height vs relative difference in percent openness.

3.2. Logistic Regression Model

The performance of logistic regression based habitat presence models indicated that LiDAR-derived parameters are efficient in RCW habitat structural assessment and for determining the roles of underlying habitat parameters. Also, in general, variation of parameter means was found to be comparatively higher in no-bird zones. On performing univariate statistical analysis, we assessed the validity of individual model variables and parameters d and h were found to have similar c-values, greater than the other two variables - p and f. Here, when the c-value nears to 1, it means that the model is capable of perfectly discriminating the response; this is equivalent to the well-known measure ROC (receiver operating characteristic) [14,17]. Correlation plots for the selected parameters revealed high association ($r = 0.71611$, $p < 0.001$) between h and d. On testing different combinations of the 4 variables, the best case (highest c-value; 0.866) was obtained for the model with d and f as independent variables. There were no statistically significant interactions found between the two variables ($p > 0.001$). p has been found to be a significant factor in similar studies [4,12], however, in our case, it had a negative impact on the c-value (0.855) and thus was removed at a later stage. Here, equation 4 represents the general form of model I; the overall model was found to be significant (Likelihood Ratio chi-square test value: 21.045; Pr < 0.001). The odds ratio estimates indicate that for every one-unit increase in d, the chances of finding RCW increased by 14 % when the effects of other factors are kept constant and the RCW favorability increased by 6 % when the f increased by one unit keeping the effects of other variables constant.

$$P = 0.1343\, d + 0.0561\, f - 8.9556 \tag{4}$$

An alternative model (model II) was built on similar procedures, though it included data from additional 20 natural sites from CNP as well. This helped us evaluate how the performance of our model was influenced by the inclusion of parameters dictating the favorability in natural habitats. In this case, univariate statistical analysis resulted in h having the highest c-value (0.763) and hence the model II had h replace d, unlike in model I. The general form of model II is listed in equation 5.

$$P = 0.0634\, h + 0.0439\, p - 7.2225 \tag{5}$$

Both the models were further evaluated based on their predictive performance; by comparing their ability to identify the presence of non-natural habitat conditions. For this validation process, 10 new non-natural study sites in Palmetto-Peartree Preserve were considered. Model I successfully identified 8 out of the 10 total sites at 75 % confidence interval, whereas model II could classify only 6 of them correctly. A 75 % confidence interval was chosen as it was reported as a decent measure in similar studies done in the ecological modeling domain.

4. Discussion

Findings from our study suggest that nesting habitats at natural and non-natural sites might have different underlying parameters contributing at varying degrees towards RCW favorability. From our results, the importance of difference in distance between the tree top and mid-story towards habitat favorability is evident and incorporating this parameter in the existing models might enable advancement of RCW conservation activities in non-natural habitat locations. The habitat model I, made of parameters extracted from 20 non-natural and 20 no-bird zones, showed f and d as dominant parameters whereas when data from additional 20 natural zones were included the primary factors changed to h and p. This in fact states that, increasing h and decreasing f results in increasing RCW presence within non-natural habitats - i.e., a positive correlation between d and habitat favorability. Also, the model II, with h and p as parameters, performed with comparatively lesser accuracy with non-natural habitat identification. This further implies that considering the exact parameters of RCW habitats in natural sites might not be able to successfully locate RCW habitat zones in non-natural habitats. Additionally, LiDAR structural assessment of the forest canopy revealed the similarity of f to the mid-story height and the logistic regression analysis results underscore the role of f in habitat prediction models. Even then, as a whole, it is difficult to affirm whether the results (8 from 10 with 75% confident interval) can prove the suitability of the proposed approach for RCW cluster mapping in non-natural sites due to limited study samples. More data collection and analysis need to be performed for bolstering our interpretations. Also, since, our major focus was on detecting non-natural habitats, and due to lack of enough high quality data of additional natural habitats at the time of study, we did not further test the applicability of models for identifying natural sites or no-bird zones. However, testing and comparing the performance of our models on variant ecosystems as well the models' compatibility with increased levels of confidence intervals are planned in upcoming studies to draw more convincing inferences. It would be also interesting to know how the habitat suitability models developed earlier would perform compared to the new models developed here using the newly found attributes. Extraction of key variables included in the earlier models from the field plot data and including those in the current models' feature selection procedures is also another prospective worth investigating; this will enable us to check if the newly identified LiDAR-features outperforms the existing ones used for model creation purposes.

In addition to RCW, this framework is transferable to several other species - such as California spotted owl (*Strix occidentalis occidentalis* Xantus de Vesey) and Fisher (*Martes pennant* Erxleben), whose habitat mapping is guided by properties of stand structure [18,19]. In these cases, field inventory methods can be outrun by LiDAR-based techniques as well as other remote sensing methods such as aerial and satellite imagery, as it allows for a vertical characterization of nesting habitat structures [10]. Multi-return high resolution LiDAR have been used for monitoring large residual trees which facilitate selection of habitats by spotted owls, and to measure number, density, pattern of trees and canopy cover for evaluating the associated nesting habitats [19]. For testing the influence of canopy cover variation in these situations, inclusion of height-density factor in the models could be found helpful.

Considering the high cost associated with LiDAR technology, and advancements in Structure-from-Motion algorithms, as a future alternative, habitat characteristics can be derived using stereo imagery obtained via Unmanned Aerial Vehicles (UAVs) as well [20,21]. This will help eradicate the limitations associated with absence of consecutive years' data; another possible way to tackle this situation is by modeling and including predictive forest growth where habitat changes can be simulated as a function of land cover change [22]. Either way, the time difference between field data collection and remotely sensed data acquisition should be proved not to be significant for the study sites (i.e., to prove that the forest stands did not change too much over the years) for providing validity to the research interpretations [23]. In this study, landscape connectedness with existing clusters was not considered; including those parameters may add more variables and/or relevant factors to the model. In that case, several areas located by our model as possible habitat zones might not be found suitable for accommodating RCW clusters if they are adjacent to existing clusters. This is because of the territorial nature of RCWs i.e., they often have individual groups that compete against each other for territory [24]. Also, it has been reported that the artificial cavities need to be at least 8.05 km (5 miles) away from the existing clusters for relocation to happen and further exhaustive research - and related model upgradations - has to be done for accurately demonstrating the effect connectivity has on habitat favorability.

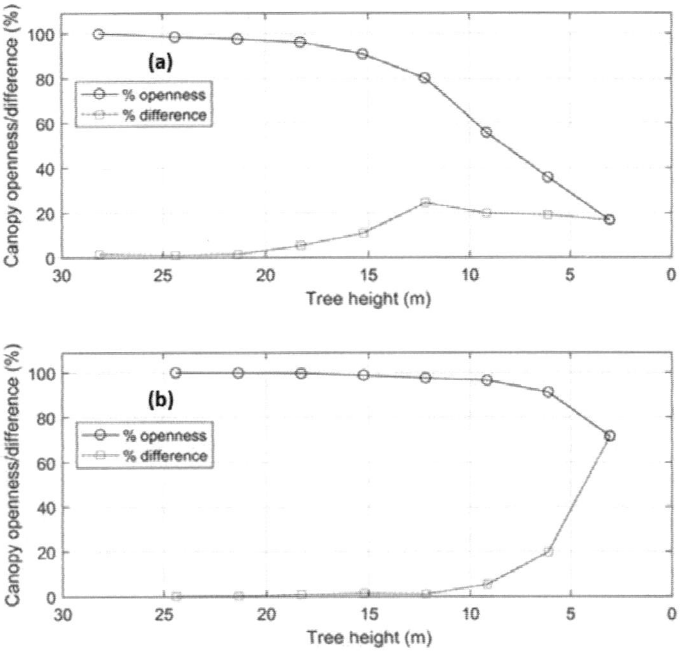

Figure 4. Change in canopy openness with varying tree height of non-natural study sites: (a) without any nearby roads (b) with a road in close proximity (< 30.48 m i.e., 100 feet) to the RCW nest cavity

Another potential parameter which should be explored is the distance between habitats and roads. In most of the existing models, we find the roads being masked [25]; nonetheless, inspecting several of our spots near roads led us to speculate that open, scarcely-used roads had acted as sparse forging zones and increased the habitat favorability, considering the change in pattern observed while calculating f for these sites (see figure 4b). Herein, using spatially explicit and detailed LiDAR-derived habitat attributes we are able to gather new insights into RCW nesting habitat relationships and this can be further underpinned by employing supplementary field plot data [26]. Also, development of thematic potential RCW presence-absence map from these kinds of models based on environmental characteristics is also a possibility; it can then be used to assess the impact of various human activities such as rising urbanization and land use change on habitat quality. For this study, we analyzed only nesting sites, however, we anticipate a similar framework to be extended to foraging habitats as well; ideally, a structural overlay of social or biological model can be developed for pursuing future research.

In sum, even though, this pilot study cannot function as a stand-alone research to provide convincing conclusions regarding why RCWs favor certain non-natural sites, for sure, the results of our study are able to develop a hypothesize that habitats that do not meet all the conditions set by well-established recovery plans can yet have the potential for nurturing healthy and advancing RCW populations. This discovery is crucial for enhancing the decision making processes associated with conservation strategies - for not just RCWs, but of similar species of birds and animals as well - in the present era influenced by rapid climate change and environmental transformation. Parameters such as f and d, which are not considered in existing models, proved to be helpful in non-natural habitat identification. Inclusion of these newly found features may offer additional options for extending the threshhold limits of pre-established features and thereby help in creation of habitat suitability indices that can account for feature deviations prominent in non-natural ecosystems. However, data acquisition from non-natural sites can be a big challenge, as these locations are not widely known. For enhancing our understanding of RCW habitat favorability and for improving the predictive range of habitat models, future studies in the direction of merging nesting habitats with foraging habits for structural analysis will be very helpful. By taking the proposed measures, we can improve the ways RCW conservation measures are integrated into the forest management methodologies and thereby create a win-win situation for both forest managers and RCW conservationists [27,28]; we also envision applications of advanced geostatistical as well as other cutting edge remote sensing techniques - such as multispectral lidar - to assist the model development initiatives [29,30,31,32]. Whilst this study focused on RCWs, we expect the proposed framework to be applicable for a wide range of endangered species and look forward to seeing climate and environment adaptable habitat models in mere future.

5. Conclusions

From this research study, we were able to hypothesize that there exist differences in values of underlying characteristics for natural and non-natural habitats, and underscores the influence height difference and height density factor have towards Red-cockaded woodpeckers' habitat favorability. However, it must be borne in mind that, since the existence of several endangered species in non-natural habitats is not widely publicized or studied, the inputs for developing respective models can be difficult to obtain. Nonetheless, by sharing our results we intend to encourage the readers, researchers and biologists to pursue studies on a larger scale in the proposed direction for determining the longer-term effects that environmental and climate changes will have on habitat favorability of various animal species. This way, effective and adaptive habitat conservation models can be developed, which can further contribute towards the protection and well-being of numerous endangered species of animals and birds.

References

[1] Dore M H 2005. Climate change and changes in global precipitation patterns: what do we know?. *Environment international*, 31(8), 1167-1181.

[2] Jackson J A 1971. The evolution, taxonomy, distribution, past populations and current status Of the red-cockaded woodpecker. Ecology and management of the red-cockaded woodpecker. *US Bureau of Sport Fishing and Wildlife and Tall Timbers Research Station, Tallahassee, FL* 1971: 4-29.

[3] Myers G 2016. *North Carolina Wildlife Resources Commission.*

[4] Lipscomb D J, Thomas M W 2006. Evaluating some proposed matrices for Scoring sub-optimal red-cockaded woodpecker foraging habitat in relation to the 2003 recovery plan. *Gen. Tech. Rep. SRS-92. Asheville, NC: US Department of Agriculture, Forest Service, Southern Research Station*. pp. 10-16.

[5] U.S. Fish and Wildlife Service; Red-cockaded Woodpecker Recovery; RCW Foraging Matrix Application; Available online: http://www.fws.gov/rcwrecovery/matrix.html (accessed on 15 December 2015)

[6] Bruggeman D J, Michael L J 2008. "Should habitat trading be based on mitigation ratios derived from landscape indices? A model-based analysis of compensatory restoration options for the red-cockaded woodpecker. *Environmental Management* 42.4 (2008): 591-602

[7] NCfloodmaps; LiDAR and Digital Elevation Data; Available online: http://www.ncfloodmaps.com/pubdocs/lidar_final_jan03.pdf (accessed on 21 October 2015)

[8] ESRI 2011. ArcGIS Desktop: Release 10. Redlands, CA: Environmental Systems Research Institute.

[9] Scott L M, Janikas M V 2010. Spatial statistics in ArcGIS. *Handbook of applied spatial analysis*, 27-41.

[10] Mohan M 2016. Spatial Modeling and Evaluation of Red-cockaded Woodpecker Non-Traditional Habitats: A Logistic Regression Analysis and LiDAR Approach. *NCSU Library.*

[11] Garabedian J E, Moorman C E, Peterson M N, Kilgo J C 2018. Relative importance of social factors, conspecific density, and forest structure on space use by the endangered Red-cockaded Woodpecker: A new consideration for habitat restoration. *The Condor*, 120(2), 305-318.

[12] U.S. Fish and Wildlife Service. 2003. Recovery plan for the redcockaded woodpecker (*Picoides borealis*): second revision. Atlanta, GA: U.S. Fish and Wildlife Service. 296 p.

[13] Pereira J, Itami R 1991. GIS-based habitat modeling using logistic multiple regression- A study of the Mt. Graham red squirrel. *Photogrammetric engineering and remote sensing*, 57(11), 1475-1486.

[14] Dayton C M 1992. Logistic regression analysis. *Stat*, 474-574

[15] R Core Team 2015. R: A Language and Environment for Statistical Computing. R Foundation for Statistical Computing, Vienna, Austria, accessed Oct. 15 2015, <http://www.R-project.org≥.

[16] Alison P D 2012. Logistic regression using SAS: Theory and application. SAS Institute.

[17] Fawcett T 2006. An introduction to ROC analysis. *Pattern recognition letters*, 27(8), 861-874.

[18] Purcell K L, Mazzoni A K, Mori S R, Boroski B B 2009. Resting structures and Resting habitat of fishers in the southern Sierra Nevada, California. *Forest Ecology and Management*, 258(12), 2696-2706.

[19] García-Feced C, Tempel D J, Kelly M 2011. LiDAR as a tool to characterize wildlife habitat: California spotted owl nesting habitat as an example. *Journal of Forestry*, 109(8), 436-443.

[20] Mohan M, Silva C A, Klauberg C, Jat P, Catts G, Cardil A, Hudak A T, Dia M 2017. Individual tree detection from unmanned aerial vehicle (UAV) derived canopy height model in an open canopy mixed conifer forest. *Forests*, 8(9), 340.

[21] Gonzalez L F, Montes G A, Puig E, Johnson S, Mengersen K, & Gaston K J 2016. Unmanned Aerial Vehicles (UAVs) and artificial intelligence revolutionizing wildlife monitoring and conservation. *Sensors*, 16(1), 97.

[22] Mueller J, Moning C, Baessler C, Heurich M, Brandl R 2009. Using airborne laser scanning to model potential abundance and assemblages of forest passerines. *Basic and Applied Ecology*, 10(7), 671-681.

[23] Silva C A, Klauberg C, Hudak A T, Vierling L A, Jaafar W S W M, Mohan M, Garcia M, Ferraz A, Cardil A, Saatchi S 2017. Predicting stem total and assortment volumes in an industrial *pinus* taeda L. Forest plantation using airborne laser scanning data and random forest. *Forests*, 8(7), 254.

[24] Hooper R G, Niles L J, Harlow R F, Wood G W 1982. Home ranges of red-cockaded woodpeckers in coastal South Carolina. *The Auk* (1982): 675-682.

[25] Copeyon C K, Walters J R, Carter III J H 1991. "Induction of red-cockaded woodpecker group formation by artificial cavity construction. *The Journal of wildlife management*: 549-675 556

[26] Garabedian J E, McGaughey R J, Reutebuch S E, Parresol B R, Kilgo J C, Moorman C E, Peterson M N 2014. Quantitative analysis of woodpecker habitat using high-Resolution airborne LiDAR estimates of forest structure and composition. *Remote Sensing of Environment*, 145, 68-80.

[27] Roise J P, Harnish K, Mohan M, Scolforo H, Chung J, Kanieski B, Catts G P, McCarter J B, Posse J, Shen T 2016. Valuation and production possibilities on a working forest using multi-objective programming, Woodstock, timber NPV, and carbon storage and sequestration. *Scandinavian Journal of Forest Research*, 31(7), 674-680.

[28] Roise J P, Chung J, Lancia R 1991. Red-cockaded woodpecker habitat management and longleaf pine straw production: an economic analysis. *Southern Journal of Applied Forestry*, 15(2), 88-92.

[29] Jat P, Serre M L 2018. A novel geostatistical approach combining Euclidean and gradual-flow

covariance models to estimate fecal coliform along the Haw and Deep rivers in North Carolina. *Stochastic Environmental Research and Risk Assessment*, 1-13.

[30] Huo L, Silva C A, Klauberg C, Mohan M, Zhao L, Tang P, Hudak A T 2018. Supervised Spatial Classification of Multispectral LiDAR Data in Urban Areas. *PLOS ONE*, 13(10): e0206185.

[31] Wan Mohd Jaafar W S, Woodhouse I H, Silva C A, Omar H, Hudak A T 2017. Modelling individual tree aboveground biomass using discrete return lidar in lowland dipterocarp forest of Malaysia. *Journal of Tropical Forest Science*, 465-484.

[32] Wan Mohd Jaafar W S, Woodhouse I H, Silva C A, Omar H, Maulud K N A, Hudak A T, Klauberg C A, Mohan M 2018. Improving Individual Tree Crown Delineation and Attributes Estimation of Tropical Forests Using Airborne LiDAR Data. *Forests*.

.

Acknowledgments

We would like to thank The Conservation Fund, North Carolina Natural Heritage Program 2016, Division of Land and Water Stewardship and Department of Natural and Cultural Resources, Raleigh, North Carolina for the use of Palmetto territory and Croatan National forest, their GIS and (Biotics) database; John Hammond, John Fennigan and U.S. Fish and Wildlife service for guidance and GIS support; Sugeerth Murugesan of Lawrence Berkeley National Laboratory (LBNL) for data visualization instructions; Stacy Nelson, Siamak Khorram, James Garribedian, Paul Taillie, Nils Peterson, Chris Moorman, Mahendra Dia and Henrique Scolforo of North Carolina State University for their comments and helpful statistical insights. This open access publication and related proceedings of the presented work were made possible with the help of UKM-YSD Fund from National University of Malaysia (UKM) (Grant number: ZF-2017-007).

IOP Conf. Series: Earth and Environmental Science **228** (2019) 012019 doi:10.1088/1755-1315/228/1/012019

Aztec shape metamaterial-based bandpass filter for C, X and Ku-band applications

M J Alam[1], E Ahamed[1], M R I Faruque[1*], M J Hossain[1], M T Islam[2]

1 Space Science Centre (ANGKASA), Institute of Climate Change (IPI), Universiti Kebangsaan Malaysia, 43600 Bangi, Selangor, Malaysia.
2 Centre of Advanced Electronic and Communication Engineering, Universiti Kebangsaan Malaysia, 43600 Bangi, Selangor, Malaysia.
rashed@ukm.edu.my

Abstract. The paper presents the structure of an Aztec shape metamaterial unit cell that is suitable for a triple band of microwave frequencies. A comparative analysis has been made on the performance of the unit cell with several array structures. A low profile dielectric substrate FR-4 is used on which the design is made. The proposed structure operates at 6.3, 8.9, 11.91, and 16.42 GHz respectively. The double negative characteristics can also be found at X and Ku-band separately. Having a compact design, double-negative characteristics and the proposed metamaterial has potential to be used for triple band applications, especially for C, X and Ku-band.

1. Introduction

Before 1968, it was quite impossible to use an engineered material for the purpose of designing. Veselago et al [1] had made it possible and shown the world the impact of the artificial material. The materials that are found in nature do not have the power to control the electromagnetic properties. However, negative permittivity (ε) and permeability (μ) in a material characteristic was a myth for everyone. Thus the use of metamaterial had come into play. The unconventional structure, power to control the electromagnetic wave beams and negative effective parameters have made the metamaterial superior to any other materials found in nature.

Metamaterial is an artificially engineered composite material that exhibits some exotic electromagnetic properties and overcomes the usual limitations of natural available materials. These extraordinary properties of metamaterial may include negative permittivity, negative permeability, inverted Snell's law and some other similar reverse electromagnetic phenomena. After the first successful invention of metamaterial by Smith and his colleagues (Smith et al. 2000), it is being utilized in almost all the applications of electromagnetism due to these exciting electromagnetic properties.

Having these unorthodox and exceptional characteristics, metamaterials can be used in cloaking [2], filters [3], design of antenna [4], super lenses [5], reduction of SAR, absorber and electromagnetic band gaps etc. In some cases, intrinsic negative permittivity is found. Different alphabetic shapes have become popular for particular operations [7]; for instance, Benosman et al. [8] introduced a double S-

Content from this work may be used under the terms of the Creative Commons Attribution 3.0 licence. Any further distribution of this work must maintain attribution to the author(s) and the title of the work, journal citation and DOI.
Published under licence by IOP Publishing Ltd

shaped metamaterial that showed negative values of η between 15.67 - 17.43GHz. Another research group proposed various U-shaped rectangular array structures left-handed aspect at approximately 5, 6 and 11GHz. A V-shaped metamaterial showed double-negative characteristic. Zhou et al. designed an S-shaped 15 × 15 mm^2 chiral metamaterial for X- and Ku-band application. Though the EMR was not higher than 4. For the purpose of application on S and C bands, Alam et al. [9] designed modified P-shaped DNG for different unit cells and array sizes. Even though multi-ban metamaterial absorbers is a promising application there have been few studies conducted focusing the design of this type of materials [10].

2. Design of the structure

The diagram of the prospective structure is itemized in Fig. 1. Here one side of the substrate is comprised of modified Aztec shape resonator with border on the edges. Each unit cell comprises with 11mm in length and 12mm in width. Copper is used as a metal element on FR4 substrate and metal element's thickness is 0.035mm and substrate thickness is 1.6mm. The dielectric constant of FR4 substrate is ε_r = 4.3, a dielectric loss-tangent is tanδ = 0.025. Designed parameters of the proposed metamaterial are enlisted in Table 1. There 'a' and 'b' are the length and width of the proposed prototypes. The specification c is the Aztec shapes one arm length and d is the length of the other arm. Moreover, 'g' and's' are same in dimension and one is the gap of split and another one is the resonator width respectively.

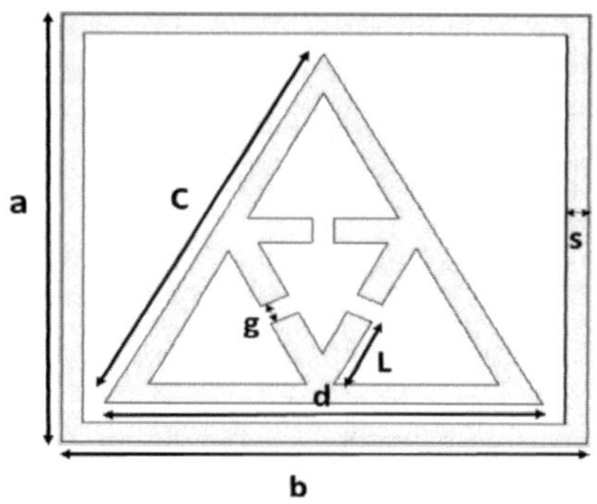

Fig. 1: Proposed Unit cell structure.

Table 1. Unit cell specification

Parameters	Dimension (mm)	Parameters	Dimension (mm)
a	11	c	10.30
b	12	d	10
s & g	0.5	L	1.48

3. Methodology

The reflection coefficient S11 named as a return loss and transmission coefficient S21 named as an insertion loss are extracted from the electromagnetic structure by using the CST Microwave Studio. The electromagnetic waves are generated using two waveguide ports and that waves are propagated through the filter configuration in Z-axis direction. The boundary condition for the metamaterial structure are applied as the perfect electric conductor (PEC) and perfect magnetic conductor (PMC) are in X- and Y- axis direction respectively. The frequency domain solver based simulator is used for free-space simulation purpose. The unit cell act like a LC resonance circuit and its resonator indicated for the circuit is inductance and its split gap behaves like a capacitor. Therefore, by applying electromagnetic field it creates extraordinary properties. By increasing resonator width or gap inductance or capacitance will be varied properly. A flexible tetrahedral mesh is used for analyzing the proposed structure. The normalized impedance is set as 50ohm and the simulation is executed from 2 to 18GHz that is shown in Fig 2. To differentiate the effective permittivity (ε_r), permeability (μ_r) and refractive index (η_r) the NRW method is used. The effective permittivity and effective permeability are determined by the equation 1 and equation 2.

$$\varepsilon_r = \frac{c}{j\pi f d} \times \frac{(1-V_1)}{(1+V_1)} \qquad (1)$$

$$\mu_r = \frac{c}{j\pi f d} \times \frac{(1-V_2)}{(1+V_2)} \qquad (2)$$

The effective refractive index (η_r) can defined by the equation 3.

$$\eta_r = \frac{c}{j\pi f d} \times \sqrt{\frac{(S_{21}-1)^2-{S_{11}}^2}{(S_{21}+1)^2-{S_{11}}^2}} \qquad (3)$$

Where, f is the frequency, $\boldsymbol{S_{21}}$ and $\boldsymbol{S_{11}}$ are the insertion loss and return loss

Fig. 2: Simulation set-up of unit cell structure.

4. Results and Discussions

4.1. Analyzing unit cell

The resonator based filter characterized by the scattering parameters. The scattering parameters reflection coefficient or return loss (S11) and transmission coefficient or insertion loss (S21) are seen in Figure 3. Unit cell, two by one array and two by two array are analyzed here for characterizing the structure. The unit cell shows pass band characteristics at 6.30 GHz, 8.90GHz, 11.91 GHz and 16.42 GHz with maximum attenuation of -12.95 dB, -12.95 dB, -18.04 dB and -28.33 dB respectively.

The effective parameters are shown in Figure 3 (b) for the simulated unit cell structure. The FR4 substrate is used to design the filter with an area of 12 × 11 mm2 for unit cell. Different types of methods are used to find out the effective parameters like Nicolson-Ross-Weir NRW methods, direct retrieval method of refractive index (DRI) methods etc. Moreover, main focused point for this work is based on the real values of the effective parameters that are retrieved from the return loss and insertion loss. All of the cases for unit cell and array structures the results are showing the negative values for the operating frequency band in case of effective parameters. From the Figure 3 and Figure 5 determined negative values for the effective parameters are almost same for the unit cell structure and array structure. The negative permittivity are shown in figure 3 at 2- 4.13 GHz (bandwidth), 6.28-7.24 GHz, 8.88-10.57 GHz, 11.93-14.27 GHz and 14.99-15.92 GHz. The permeability shows negativity at 2-4.13 GHz, 8.86-15.40 GHz and 17.69- 18 GHz. The refractive index shows negativity at 7.48-8.93 GHz, 9.20-12 GHz and 12.28-16.34 GHz. And it shows double negative properties at 9.26 GHz with negative permittivity value -1.89, negative permeability value -24.94 and negative refractive index value -115.47.

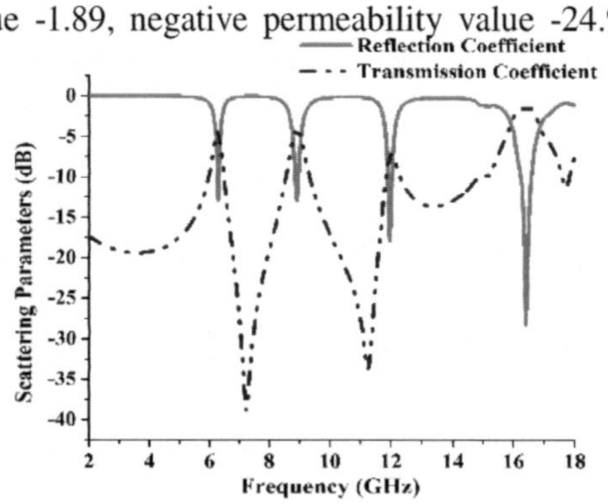

(a)

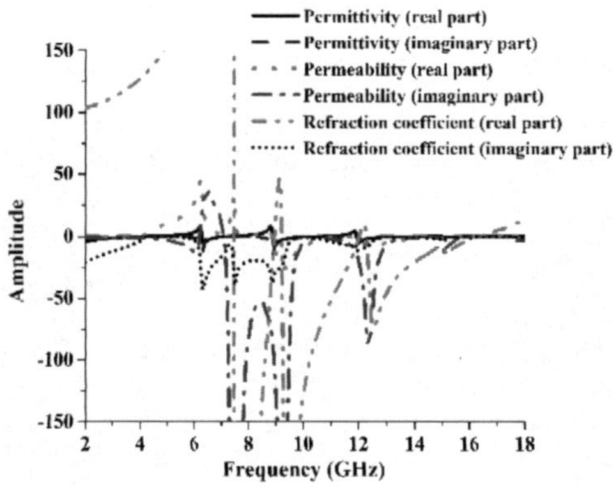

(b)

Fig. 3: Unit cell based simulated (a) return loss and insertion loss & (b) effective parameters (permittivity, permeability and refractive index) of the pass band filter.

Figure 4 indicate the situation of the proposed structure when electromagnetic field is applied along with electric field and magnetic field. Its electric field is shown in figure 4(a) where its maximum induced field indicated in the inner resonator's split gaps. Figure 4(b) represent the magnetic field where it induced in the outer resonator and figure 4(c) indicate the surface current distribution of the proposed passband structure.

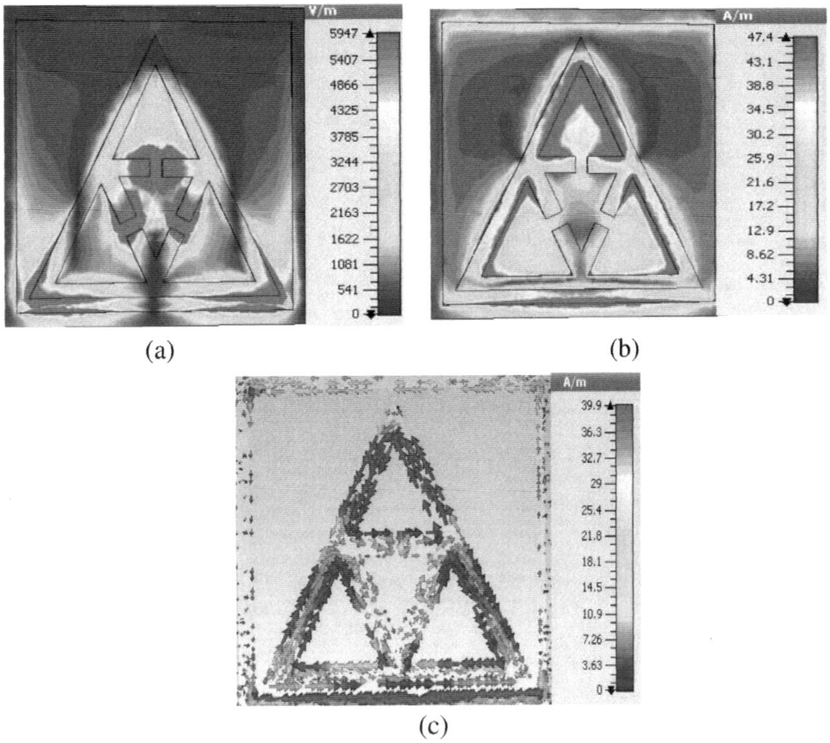

Fig. 4: Unit cell (a) electric field, (b) magnetic field & (c) surface current distribution at 9.26 GHz

4.2. Analyzing 2×1 array

The scattering parameters reflection coefficient or return loss (S11) and transmission coefficient or insertion loss (S21) are seen in figure 4. The two by one array structure shows the return loss resonances are at 6.33 GHz, 8.92GHz, 12.50 GHz and 16.52 GHz with maximum attenuation of -9.95 dB, -24.04 dB, -14.13 dB and -24.91 dB respectively. The negative permittivity are shown in figure 4 at 2- 4.16 GHz, 6.34-7.26 GHz, 8.92-10.71 GHz and 12.42-15.04 GHz. The permeability shows negativity at 7.49- 16.36 GHz. The refractive index shows negativity at 2-4.54 GHz, 8.90-12.16 GHz, 12.33- 14.31 GHz and 12.28-16.34 GHz. And it shows double negative properties at 9.26 GHz with negative permittivity value -5.44, negative permeability value -71.04 and negative refractive index value -23.53.

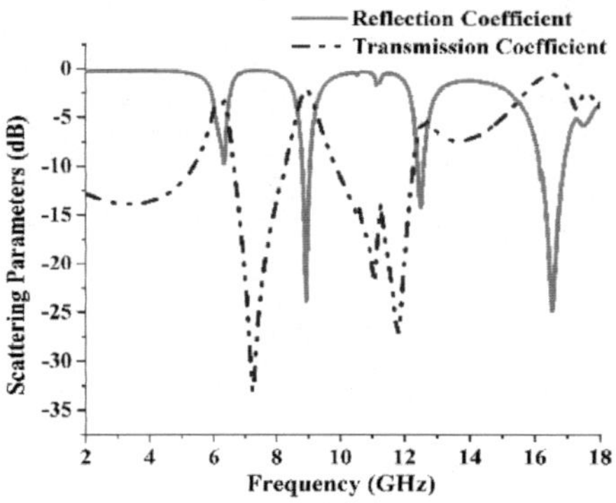

(a)

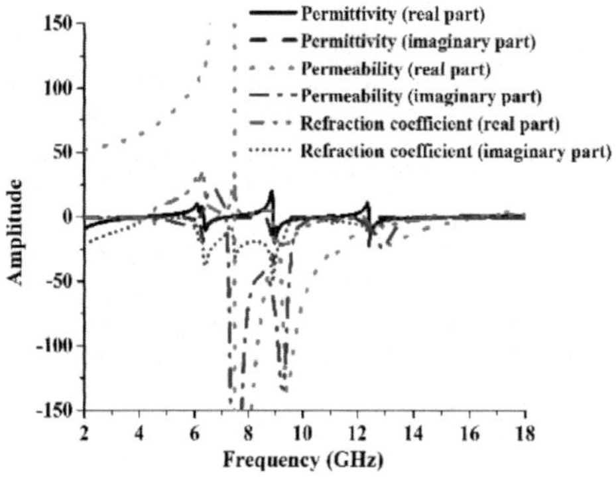

(b)

Fig. 5: 2×1 array based simulated (a) return loss and insertion loss & (b) effective parameters (permittivity, permeability and refractive index) of the passband filter.

Figure 6(a), 6(b) and 6(c) represent the electric field, magnetic field and surface current distribution at 9.26 GHz for the proposed 2×1 array structure. Where the electric field maximum concentrate in split gaps of Aztec structure and magnetic field induced in outer resonator of the structure. In the surface current distribution, currents are flowing in opposite directions between two resonators. So they cancel each other's effect and creating resonance frequency.

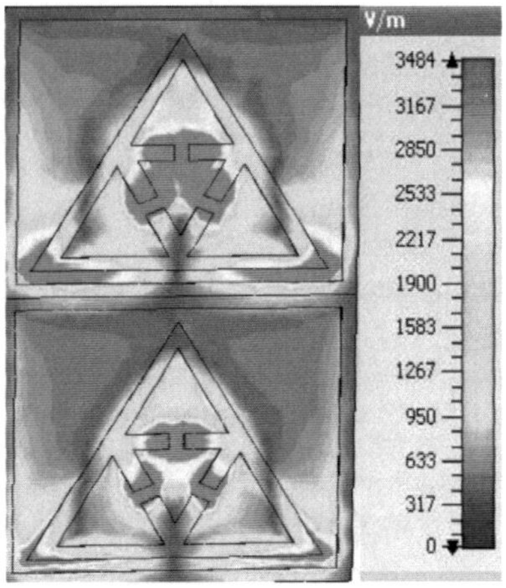

(a)

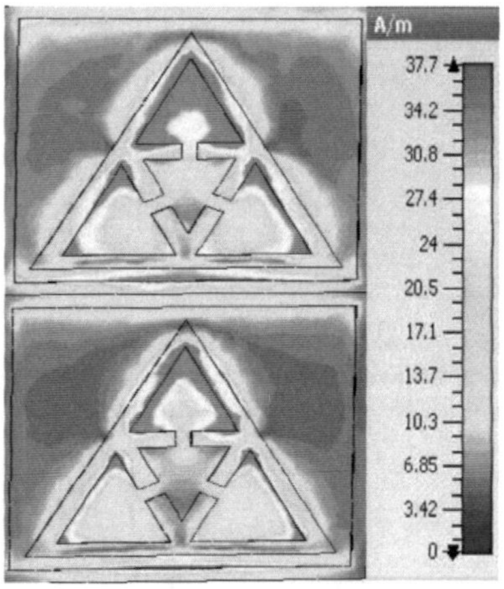

(b)

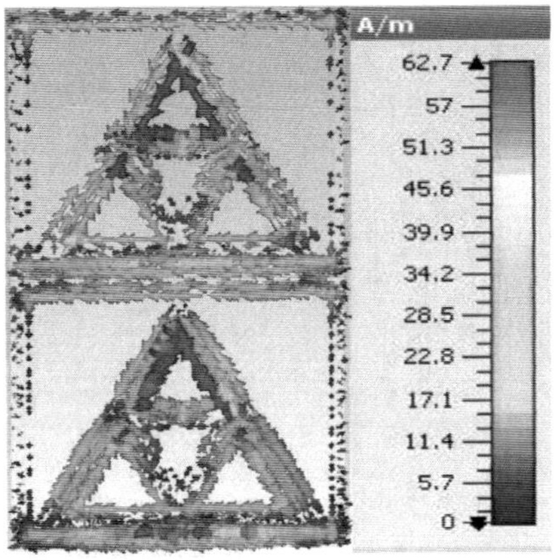

(c)

Fig. 6: 2×1 array (a) electric field, (b) magnetic field & (c) surface current distribution at 9.26 GHz

4.3. Analyzing 2×2 array

The scattering parameters reflection coefficient or return loss (S11) and transmission coefficient or insertion loss (S21) are seen in figure 5. The two by two array structure shows the return loss resonances are at 6.33 GHz, 8.92GHz, 12.44 GHz and 16.40 GHz with maximum attenuation of -15.55 dB, -12.57 dB, -21.95 dB and -37.71 dB respectively. The negative permittivity are shown in figure 3 at 2- 4.16 GHz (bandwidth), 6.34-7.28 GHz, 8.92-10.70 GHz and 12.41-15.13 GHz. The permeability shows negativity at 7.49-8.99 GHz, 9.24-12.42 GHz and 12.82-16.36 GHz. The refractive index shows negativity at 2-4.13 GHz, 8.90-14.86 GHz and 15.9-16.27 GHz. And it shows double negative properties at 9.26 GHz with negative permittivity value -2.70, negative permeability value -34.45 and negative refractive index value -20.92.

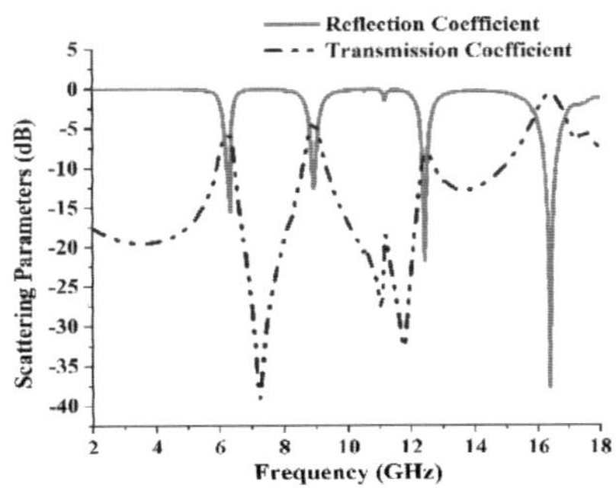

(a)

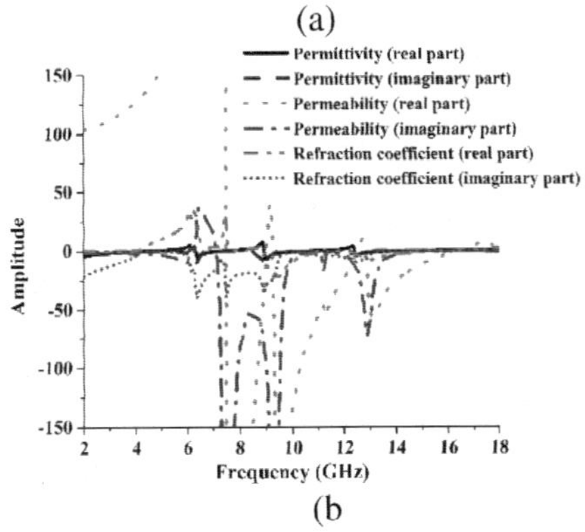

(b

Fig. 7: 2×2 array based simulated (a) return loss and insertion loss & (b) effective parameters (permittivity, permeability and refractive index) of the passband filter

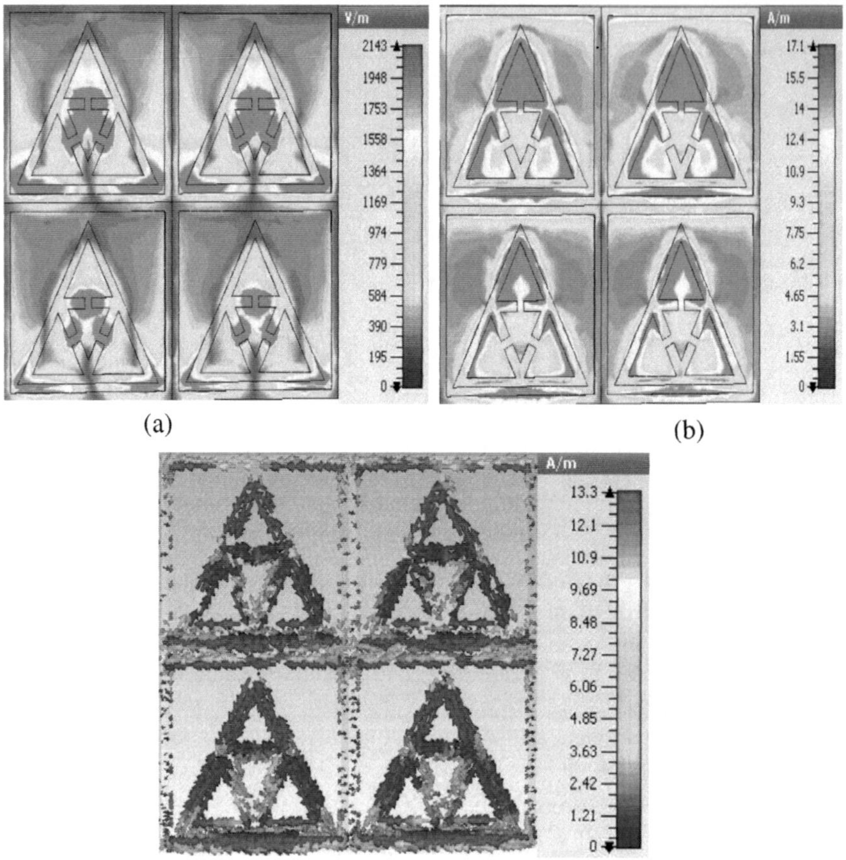

(a) (b)

(c)

Fig. 8: 2×2 array (a) electric field, (b) magnetic field & (c) surface current distribution at 9.26 GHz.

Figure 8(a), 8(b) & 8(c) indicates the electric field, magnetic field and surface current distribution of the Aztec 2×2 array structure. Here also electric field induced in the split gaps of the inner structure and magnetic field induced in the outer resonator. The magnetic field depends on the current therefore, where the magnetic field induced concentration is maximum there the surface current concentration is high indicated by red colour in figure 8(c).

5. Conclusion

The proposed Aztec shaped pass band filter presents here with its different framework. The borders are embedded around the unit cell structure to create steadier operation through the whole unit cell at 6.30 GHz, 8.90GHz, 11.91 GHz and 16.42 GHz, respectively. The performance of the proposed filter structure is verified by its return loss, insertion loss, relative permittivity, and permeability. The overall performance and gain from the proposed filter is analysed analytically. Analytical resonating points are performed well by the effective parameters for the Aztec shaped filter. Its compact size, cost effectiveness and operative frequency range made that structure applicable. The proposed structure is potentially used as a band pass filter because of its appropriate architecture and its double negative characteristics.

6. References

[1] Veselago V G 1968 *Sov Phys.* **10** 509.
[2] Faruque M R I and Islam M T 2012 *Prog. Electromagn. Res.* **124** 119.
[3] Islam M M, Islam M T, Samsuzzaman M and Faruque, M R I 2015 *Electron. Lett.*, **51** 1222.
[4] Hasan M M, Faruque M R I. and Islam M T 2018 *Scientific Reports.* **8** 1-17.
[5] Alam M J, Faruque M R I and Islam M T 2018 *J. Phys. D: App. Phys.* **51** 265102.
[6] Islam S S, Faruque M R I. and Islam M T 2015 *AIP Advance.* **5** 107116.
[7] Song K, Fu Q and Zhao X U 2011 *Phys Sci.* **84** 035402.
[8] Huang Y, Wen J, Yang Y and Xie K *Appl Phys A.* **106** 79.
[9] Benosman H and Hacene N B 2012 *Int J Comput Sci.* **9** 534.
[10] Alam M J, Faruque M R I, Hossain M J and Islam M T 2018. *Mic. Opt. Tech. Lett.* **60** 1388.

IOP Conf. Series: Earth and Environmental Science **228** (2019) 012020 doi:10.1088/1755-1315/228/1/012020

Design and analysis of compact perfect metamaterial absorber for X-band applications

M J Hossain[1], M R I Faruque[1], M R Ahmed[2], S Abdullah[1], M T Islam[3]

[1] Institute of Climate Change (IPI), Universiti Kebangsaan Malaysia(UKM) Bangi, Malaysia
[2] Electrical and Electronic Engineering, Dhaka University of Engineering and Technology (DUET), Gazipur, Bangladesh
[3] Centre of Advanced Electronic & Communication Engineering, Universiti Kebangsaan Malaysia(UKM), Bangi, Malaysia

rashed@ ukm.edu.my

Abstract. A new tri-ring hexagonal with meandered line design structure based on simple outline is proposed for X-band application in microwave ranges. The finite-difference time-domain (FDTD) method-based CST simulator was approved to observe the perfect metamaterial absorbance (PMA) analysis. Analysis was completed by using three configurations of PMA such as 1 × 1 unit-cell structure; 1 × 2 unit-cell and 1 × 2 array unit-cell structures; 2 × 1 unit-cell and 2 × 1 array unit-cell structures. The structure revealed the resonance frequency within the X-band of the microwave spectra. At resonance frequencies, the absorbance's peaks are 99.95% at 8.006 GHz and 99.61% at 11.525 GHz, correspondingly. X-band operating frequency has been obtained by all design configurations. The dimension of the offered structure is 10.50 mm × 10 mm that comprises all geometrical parameters to suitable the structure inside the area of substrate material. The results of the offered PMA exhibited X-band absorber response over the frequency ranges from 5 to 14 GHz. Hence, the offered structure enables numerous application areas, for instance, satellite communications, security, protection, stealth innovation and wireless computer networks.

1. Introduction

Negative refractive index metamaterials are engineered materials with negative permittivity and permeability simultaneously, which have exotic properties in terms of well-known phenomena like Snell's law, and Doppler Effect etc. Due to the interesting properties, the scientific community has been concerned to the metamaterials for strange characteristics and extensive applications in energy harvesting, invisibility clocks, filtering, electromagnetic absorber [1] and perfect lenses, SAR reduction [2]. Kollatou et al. offered a polarization-insensitive metamaterial absorber (MA) which worked in the microwave spectra. They gained the maximum absorbance peak of 95.81% at 10.31 GHz [3]. In this article, authors achieved very low absorbance rate at resonance frequency of 10.31 GHz instead of small design structure. Landy et al. proposed the metamaterial design structure for absorbing electromagnetic wave with an absorbance of 96% [4]. On the other hand, the experimental result of the absorber was 88% at 11.50 GHz frequency. From this article, it was shown that they used cut wire in the back side of the design structure. For that reason, the transmission parameter of the incident electromagnetic wave was not fully zero. Moreover, they got the less absorbance from their design structure by numerically and experimentally because of cut wire of the back side of the unit

Content from this work may be used under the terms of the Creative Commons Attribution 3.0 licence. Any further distribution of this work must maintain attribution to the author(s) and the title of the work, journal citation and DOI.
Published under licence by IOP Publishing Ltd

cell. Mohammad et al. proposed a Jerusalem cross with meandered load absorber which has a dimension of 24×24 mm^2 exhibits more absorption like 95% [5]. Author proposed very large size of the design structure and achieved very low absorbance rate of 95%. Lin et al. suggested a design structure that has a dimension of 1.092 cm ×1.0 cm. It was appropriate in microwave spectra. The structure depicted the peaks of absorbance at different resonance frequencies such as 96.5% at 2.15 GHz, 96.8% at 2.28 GHz, and 99.6% at 2.38 GHz, correspondingly [6]. Authors obtained multi absorbance rate at multi resonance frequency, however, they got less absorbance rate at first two resonance frequencies. An ultra-broadband perfect absorber with 10×10 mm^2 dimensions depends on electric resonator proposed by Zhao et al. The absorption peak of 99.3% at 5.45 GHz, 97.1% at 15.46 GHz and 98.6% 19.48 GHz, simultaneously [7]. Authors also achieved lower absorbance rate at higher frequencies such as 15.46 GHz and 19.48 GHz, whereas the size of the design structure was little bit compact. A unity absorbance metamaterial was designed and fabricated by Dincer et al. [8]. They depicted a PMA, which have absorbance pinnacles of 99.99% and 99.92% at 5.48 GHz 0.865 THz, correspondingly. Nevertheless, the size of the design structure was 3.6 cm × 3.6 cm.

In this paper, the both side planar multi-ring hexagonal with meandered line-based metamaterial absorber depicts dual resonances X-bands with different structural analysis such as 1 × 1 unit-cell structure; 1 × 2 unit-cell and 1 × 2 array unit-cell structures; 2 × 1 unit-cell and 2 × 1 array unit-cell structures. In addition, the dimension of the offered structure is 10.5 mm × 10 mm × 1.6 mm, which is smaller than the proposed metamaterials structure in [8, 9, 10, 12]. Also, the plan structure offers high absorbance value 99.95% at 8.006 GHz and 99.61% at 11.525 GHz, correspondingly. It is perceived that the absorbance of the offered structure is superior to the proposed MA structure in [3, 4, 5, 6, 7, 9, 10, 11, 12]. To govern the absorbance properties, the CST studio simulator 2015 was utilized.

2. Design structure and numerical simulation

The suggested structure consists of tri-ring hexagonal with meandered line resonators which was employed to attain unconventional characteristics of metamaterials that were usually not found in nature. The offered metamaterial absorber structure and parameter specification are illustrated in Figure 1(a). Figure 1(b) indicates the numerical simulation setup of proposed structure. FR4 substrate material was utilized as a separator between the resonator and the ground plane. The resonators and ground plane are built by copper material thickness of 35μm. The conductivity of resonators and ground plane are 5.8×10^7 S/m. The loss tangent δ and standard relative permittivity are 0.025 and 4.3. The specifications of the structure are Ws=10.5 mm, Ls=10 mm, W_1=0.4 mm, W_2=0.65 mm, and W_3=0.7 mm.

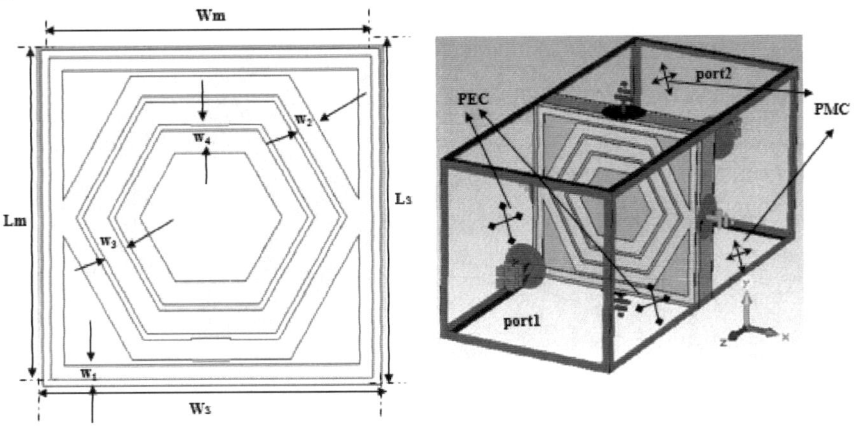

(a) (b)

Figure 1. (a) The offered absorber structure (b) Simulation setup with boundary condition.

In this paper, the FDTD method based CST simulator was applied to observe this design structure. This method shows more accurate result comparing the other numerical tools like High Frequency Structure Simulator (HFSS), IE3D, etc. It can be used to analysis the design structure very easily. As a boundary condition, two waveguide ports are setup along the z-direction. On the other hand, along the y-direction and x-direction, the perfect magnetic conductor and perfect electric conductor were applied in numerical simulation. Frequency domain solver of CST simulator was applied to find out the transmittance and the reflectance in numerical simulation and frequency samples of 1001 were considered. The absorption of the unit-cell structure was governed by the following formula, $A(\omega) = 1 - R(\omega) - T(\omega)$, where $T(\omega)$, $R(\omega)$, and $A(\omega)$ are the transmittance, reflectance and absorbance, individually. Absorption be governed by on the scattering parameters, for example, $|S_{21}|^2 = T(\omega)$, and $|S_{11}|^2 = R(\omega)$. Because of ground plane, the transmittance of the structure is zero means $T(\omega) = 0$. Therefore, from the absorbance formula, it has to be realized that decreasing the reflectance parameters which has to expand the absorbance of the metamaterial absorber.

3. Results and discussion

The reflectance ($R(\omega)$) and absorbance ($A(\omega)$) of the PMA are demonstrated in Figure 2. The deep values of the reflectance at resonance frequencies are -33.334 dB at 8.006 GHz and -23.626 dB at 11.525 GHz, respectively. Then again, the pinnacle estimations of absorbance are 99.95% at 8.006 GHz and 99.61% at 11.525 GHz, individually. It is seen from Figure 2 that less value of reflectance creates more value of absorbance.

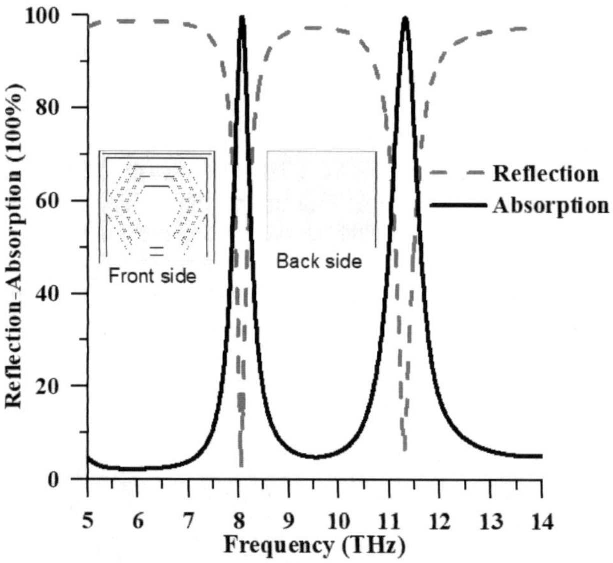

Figure 2. Reflectance and absorbance of unit-cell structure.

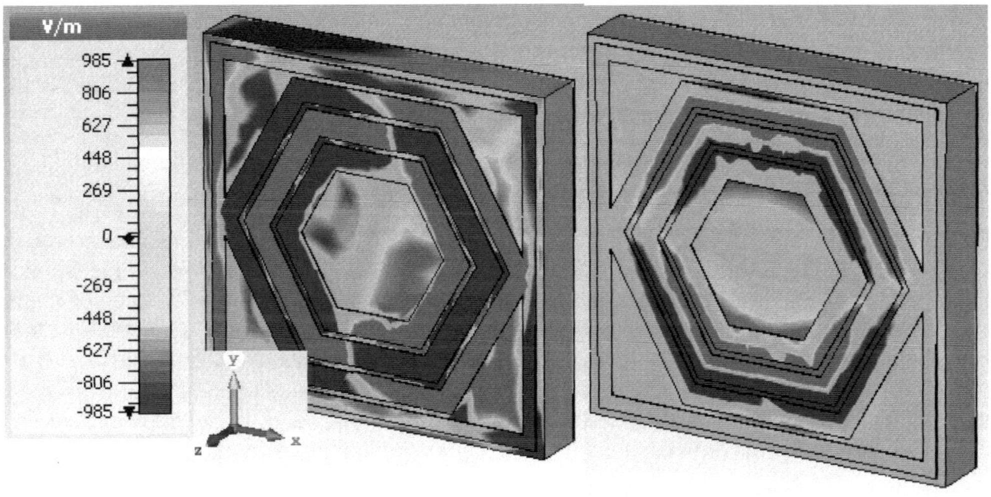

(a) (b)

Figure 3. Electric field distribution (a) at 8.006 GHz, (b) at 11.525 GHz of the structure.

Figure 3 depicts the electric field of the resonator with two peaks such as 8.006 GHz, and 11.525 GHz, where its expose high absorbance peak respectively. At 8.006 GHz, the distributions of electric field have concentrated on the areas of outer ring and inner ring of right side. It is also seen around the left side of the middle ring. On the other hand, at 11.525 GHz, the concentration of electric field present between the gap of upper and middle ring on the upper side. Hence, the high absorption peak is subjected due to the electric resonance of the regions.

3.1 Analysis of 1 × 2 unit-cell structure

The fundamental structure of 1 × 2 unit-cell is composed on the separate substrate materials. Two structures of the unit-cells are horizontally detached by 0.4 mm and 5-14 GHz frequency ranges are used in numerical simulation for analysis this design structure.

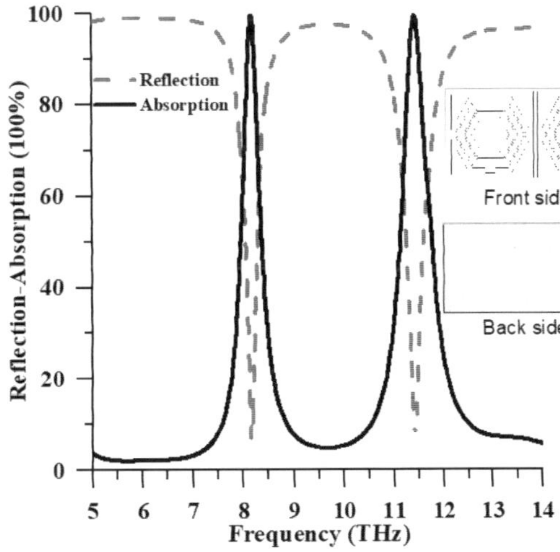

Figure 4. Reflectance and absorbance of 1 × 2 unit-cell structure.

The reflectance (R(ω)) and absorbance (A(ω)) of the PMA are presented in Figure 4. The negative peak values of the reflectance of resonance frequencies are -33.04 dB at 8.15 GHz and -22.41 dB at

11.44 GHz, separately. However, the absorbance's pinnacle estimations 99.95% and 99.48%, at 8.15 GHz, 11.44 GHz, correspondingly. It is seen from Fig. 4 that the peak of absorbance increased of first resonance, on the other hand, second absorbance peak decreased due to joint of the two unit-cells separately in the 1 × 2 unit-cell horizontal form.

3.2 Analysis of 1 × 2 array unit-cell structure

The fundamental structure of 1 × 2 array unit-cell outlined on a similar substrate material. Two front side resonators are horizontally detached by 0.4 mm and 5-14 GHz frequency ranges are utilized in numerical simulation for analysis this design structure. Figure 5 indicates that the reflectance ($R(\omega)$)and absorbance ($A(\omega)$)of the design structure. The deep values of the reflectance of resonance frequencies are -24.68 dB at 8.17 GHz and -23.08 dB at 11.41 GHz, separately. Nonetheless, the peak values of absorbance at 8.17 GHz be 99.66% and at 11.41 GHz be 99.51%, respectively. It is seen from Figure 5 that the peak of absorbance of first resonance is increased and shifted toward the upper frequency, though, second resonance is decreased toward lower frequencies with lower peak in order to joint of the two unit-cells same substrate and same ground plane in the 1 × 2 array unit-cell horizontal form.

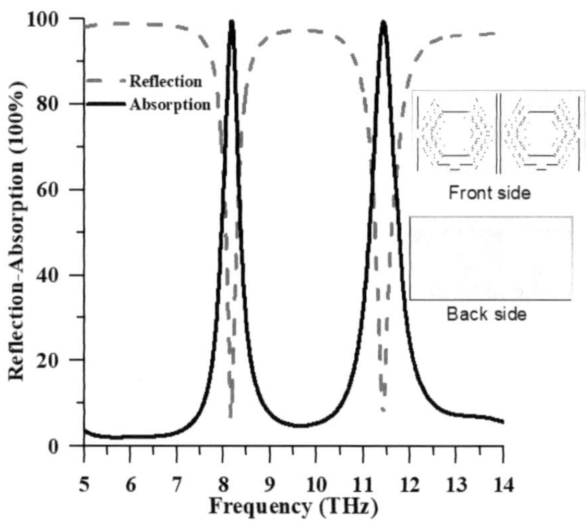

Figure 5. Reflectance and absorbance of 1 x 2 array unit-cell structure.

3.3 Analysis of 2 × 1 unit-cell structure

The basic structure of 2 × 1 unit-cell composed on the separate substrate materials. Two structures of unit-cell are vertically detached by 0.4 mm and 5-14 GHz frequency ranges are used in numerical simulation for analysis this design structure. The reflectance ($\boldsymbol{R(\omega)}$)and absorbance ($\boldsymbol{A(\omega)}$)of the PMA are display in Figure 6. The negative peak values of the reflectance of resonance frequencies are -20.50 dB at 8.13 GHz and -22.56 dB at 11.53 GHz, separately. However, the absorbance's pinnacle estimations 99.11% and 99.47%, at 8.13 GHz, 11.53 GHz, correspondingly. It is seen from Fig. 6 that the peaks of absorbance of first and second resonances are decreased compared with fundamental structure due to joint of the two unit-cells separately in the 2 × 1 unit-cell vertical form.

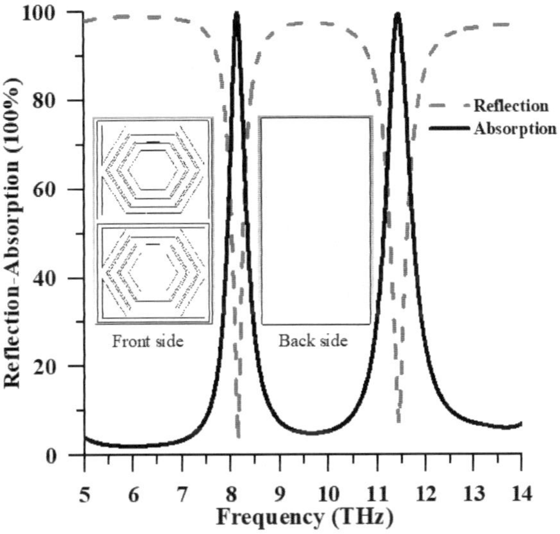

Figure 6. Reflectance and absorbance of 2 × 1 unit-cell structure.

3.4 Analysis of 2 × 1 array unit-cell structure

The basic structure of 2 × 1 array unit-cell is outlined on a vertically similar substrate and ground plane. Two structure unit-cells are vertically detached by 0.4 mm and 5-14 GHz frequency ranges are applied in numerical simulation to analysis this design structure. The reflectance ($\boldsymbol{R}(\boldsymbol{\omega})$)and absorbance ($\boldsymbol{A}(\boldsymbol{\omega})$)of the design structure are illustrated in Figure 7. The deep values of the reflectance of resonance frequencies are -22.35 dB at 8.13 GHz and -26.42 dB at 11.16 GHz, correspondingly. On the other hand, the absorbance's pinnacle estimations are 99.42% and 99.76%, at 8.13 GHz, 11.42 GHz, correspondingly. It is observed from Figure 7 that the peak of absorbance increased both of the resonances and shifted toward the lower frequencies due to joint of the two unit-cells separately at same substrate and the ground plane in the 2 × 1 unit-cell vertical form.

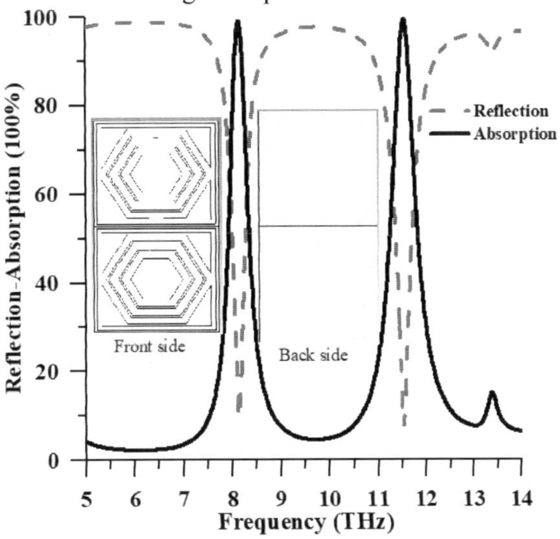

Figure 7. Reflectance and absorbance of 2 × 1 array unit-cell structure.

The assessments with offered PMA and previous PMA. Here, the parameters of the comparisons are structured shape, covered band, dimension, maximum absorbance rate, author name and published year.

The dimension (36×36 mm^2) of square shape is suggested by Dincer et al. [8], whereas authors achieved 99.99% maximum absorbance peak in C-band application.

Table 1. Comparison among offered PMA with previous PMAs

Author name with reference	Structured shape	Covered band	Size (mm^2)	Maxi^m peak of absorbance (%)	Published year
Dincer et al. [8]	Square shape	C-band	36 × 36	99.99	2015
Rana et al. [9]	U-shape	X-band	15 × 15	98.00	2016
Borah et al. [10]	O-shape	X-band	12 × 12	98.90	2016
Sen et al. [11]	L-shape	X-, Ku-band	9 × 9	95.00	2017
Mahmood et al. [12]	Modified S-shape	X-band	16 × 16	90.00	2017
This paper	Hexagonal shape	X-band	10.5 × 10	99.95	---

In addition, a U-shape multiband absorber designed by Rana et al. that had a dimension (15×15× mm^2). Authors got the peak of absorbance (98%) with big design structure. Sen et al. suggested a compact design structure (9×9 mm^2) compare with offered structure (10.5×10 mm^2). However, offered metamaterial absorber gained high absorbance peaks (95.00%) that are shown in Table1. A large size, design structure (12×12 mm^2) was suggested by Borah et al. in this paper. The authors achieved less absorbance about 98.90% compared with offered PMA. Mahmood et al. recommended a big size (16×16 mm^2) unit-cell structure, whereas they had to obtain less absorbance rate (90.00%) comparison with offered design structures. A new compact (10.5×10 mm^2) multi-ring hexagonal with meandered line design structure analyzed and achieved a high absorbance rate (99.50%) in this study. The offered metamaterial absorber achieved high absorbance and compactness comparing stated references which are appropriate for the microwave spectra.

4. Conclusions

A new design of PMA was offered for mono-band, for example, X-band applications in the manuscript. The design unit-cell structure and different array structure were analyzed for absorbing purposes over a frequency range of 5-14 GHz. Seven structural designs such as 1 × 1 unit-cell structure; 1 × 2 unit-cell and 1 × 2 array unit-cell structures; 2 × 1 unit-cell and 2 × 1 array unit-cell structures are designed and analyzed to observe absorbance. The electric field distributions also analyzed to observe the behavior of offered structures. The CST electromagnetic simulator was utilized to find out the absorption for different structures. The offered PMA is proper for X-band applications. A comparative examination additionally was done based on the author name with reference, structure shape, covered band, size, maximum absorbance rate, and published year. It was observed that the maximum absorbance rate of 99.95% at 8.006 GHz was attained from the 1×1 unit-cell structure. Hence, the PMA was compact in dimension and high absorbance which were more appropriate in microwave bands.

References

[1] Bowen P T, Baron A and Smith D R 2016 Theory of patch-antenna metamaterial perfect absorbers *Phys. Rev. A* **93** 063849
[2] Kwak S il, Sim D-U, Kwon J H and Yoon Y J 2017 Design of PIFA with metamaterials for body-SAR reduction in wearable applications *IEEE Trans. Electromagn. Comp.* **59** 297-300
[3] Kollatou T M, Dimitriadis A I, Assimonis S D, Kantartzis N V and Antonopoulos C S 2013 A family of ultra-thin, polarization-insensitive, multi-band, highly absorbing metamaterial

structures *Prog. Electromagn. Res.* **136** 579–94

[4] Landy N I, Sajuyigbe S, Mock J J, Smith D R and Padilla W J 2008 Perfect metamaterial absorber *Phys. Rev. Lett.* **100** 207402–04

[5] Naser-Moghadasi M, Nia A Z, Toolabi M and Heydari S 2017 Microwave metamaterial Absorber based on Jerusalem Cross with meandered load for bandwidth enhancement *Opt. Int. J. Light Electron Opt.* **140** 515–22

[6] Lin B, Zhao S, Da X, Fang Y, Ma J, Li W, Zhu Z 2015 Triple-band low frequency ultra-compact metamaterial Absorber *J. Appl. Phys.* **117** 184503:1-5

[7] Zhao J, Cheng Y 2016 Ultrabroadband microwave metamaterial absorber based on electric SRR loaded with lumped resistors *J. Electron. Mater.* **45** 5033–39

[8] Dincer F, Karaaslan M, Sabah C 2015 Design and analysis of perfect metamaterial absorber in GHz and THz Frequencies *J. Electromagn. Waves Appl.* **29** 2492–2500

[9] Rana C R, Inamdar K, Lalluwadia 2016 A Study, Design and Analysis of U-Shaped Metamaterial Absorber for X-Band Application *IUP J. Telecommun.* **8** 23-37

[10] Borah D, Bhattacharyya N S 2017 Design and Development of Expanded Graphite-Based Non-metallic and Flexible Metamaterial Absorber for X-band Applications *J. Electron. Mater.* **46** 226–232

[11] Sen G, Islam S N, Banerjee A, Das S 2017 Broadband Perfect Metamaterial Absorber on Thin Substrate for X-Band and Ku-Band Applications *Prog. Electromagn. Res. C* **73** 9–16

[12] Mahmood A, Yetkin G Ö, Sabah C 2017 Design and Fabrication of a Novel Wideband DNG Metamaterial with the Absorber Application in Microwave X-Band *Adv. Condens. Matter Phys.* **2017** 1279849:1-8

Sustainable assessment of academic buildings at Universiti Teknologi Petronas campus

W A Al Madhoun[1], A A Firdaus[1] and H A Isiyaka[2]

[1]Civil and Environmental Engineering Department, Universiti Teknologi Petronas, 32610 Seri Iskandar, Perak Darul Ridzuan, Malaysia.
[2] Department of Geography, Faculty of Earth and Environmental Sciences, Kano University of Science and Technology Wudil, Nigeria.

Abstract. The objectives of this study are; to investigate the most suitable method for Sustainable Building Assessment, to gather the data of specific indicator and validate each criteria according to sustainable building assessment benchmark. However, variables like energy efficiency, indoor environmental quality, sustainable site planning and management, available materials and resources, water efficiency and innovation were integrated and separately studied to develop a sustainable academic building environment in the campus of university Petronas, Malaysia. Sustainable assessment tools, GBI tool and in-depth literature review were used. The result shows that GBI sustainable building assessment of Universiti Teknologi Petronas scores 43 out of 64 which is 67% of the total score hence the rating would be Silver. This result proves that some improvements need to be made on some parameters, which did not pass the requirements. In a nutshell, Universiti Teknologi Petronas has high potential to meet GBI certified level that will help to increase the ranking status of the university to higher level. This also would be a great opportunity for the university to set a benchmark as a role model of green university in Malaysia. However, this study recommends that UTP management develops corrective policies to improve the building overall performance and standard.

1. Introduction
The energy used in construction, occupying and operating buildings is the major contributor to climate change that has various impacts on the environment. Therefore, constructions and building industries have the potential to recover the natural environment and community culture by conforming to the sustainable benchmarks set by the local and global requirements while enhancing the value of products and services to business, customers and society in general. To comply with these needs, there has been an increasing interest in sustainable assessments of the built environment. Recently, we can now find few number of great sustainable assessment tools regardless of new or built environment, focusing on energy use in buildings, building materials, optimized operational, conservation of water resource,

Content from this work may be used under the terms of the Creative Commons Attribution 3.0 licence. Any further distribution of this work must maintain attribution to the author(s) and the title of the work, journal citation and DOI.
Published under licence by IOP Publishing Ltd

preservation of regional and cultural identity, convenient indoor climate and maintenance practices. Universities need to be sustainable in every attribute, economic, social and environment. Therefore, Sustainable Building Assessment has to be done to evaluate and find the key indicator or criteria required to measure in order to rank the university campus sustainability. By investigating different methods of assessment, this study would be able to explore the suitability and adaptability of the tools by comparing the criteria and benchmarks to Malaysia Practice. From here, the method could be implemented to assess the building sustainability of Universiti Teknologi Petronas building campus. The objectives of this study are; to investigate the most suitable method for Sustainable Building Assessment, to gather the data of specific indicator or criteria, to analyse and validate each criterion according to sustainable building assessment benchmark.

2.0 Literature Review

Buildings can be classified as sustainable when all the sustainability aspects such as environmental, economic, social and cultural are evident in the buildings. The main reasons to achieve sustainable buildings are to solve the issues and concerns regarding wastes, pollutants, non-renewable materials, water and reduction of harmful emissions. To comply with these key aspects, several methods of building sustainability assessments have been developed with same goal which is to achieve the optimisation of site potential, preservation of regional and cultural identity, minimisation of energy consumption, protection and conservation of water resources, use of environmentally friendly materials and products, a healthy and convenient indoor climate, and optimised operational and maintenance practices.

2.1Development of Sustainable Assessment Tool

The main objective of sustainability assessment is to collect and report data for decision-making during different phases of the construction, design and use of building [1]. The development of assessment methods and respective tools are very complicated to practice and understand due to major issue of managing the flow of information and knowledge between the various level of indicator systems [1]. There are many Sustainable Building Assessment methods currently available in the market and they are widely used in the environmental product declaration. These methods have been developed worldwide and are built upon various principles and different evaluation items, data and criteria. However, a comparison of sustainable assessment tool is described in table 1.

2.2 Sustainable Assessment Tool Comparison

Table 1: Sustainable Building Rating Tool Comparison

Comparison Item	Sustainable Assessment Tools			
	BREEAM	SB TOOL	LEED	GBI
Flexibility	Moderate	High	Moderate	High
Market	Strong	Moderate	Strong	Moderate
Characteristic	Two Process	4 Phases	Industrial Standard	Tropical Climate

2.3 The Implementation of GBI Tool

The purpose of GBI is to enable green grading and certification of Malaysian buildings. GBI is a rating system providing a comprehensive framework for building assessment, which is similar to BREEAM (UK), LEED (USA), Green Star (Australia) and Green Mark (Singapore). [2] states that Malaysia's GBI focuses on indoor environmental quality, sustainable site planning and management, materials and resources, water efficiency and innovation. GBI has simplified and standardized the green building concept in Malaysia so that no misunderstanding occurs. GBI define green buildings by establishing a common language and standard of measurement. Other than that, it promotes integrated whole building designs that provide a better environment. In addition, they recognize and reward environmental leadership and transform the built environment to reduce its negative environmental impact. Lastly, GBI ensures new buildings remain relevant in the future and existing buildings are refurbished and upgraded to non-governmental and profession driven green rating tool developed for the tropical climate [2].

3.0 Methodology

This study was carried out in Universiti Teknologi Petronas academic buildings where several data were collected from authorities of both management and technical departments in order to fulfil the data requirement for Sustainable Building Assessment. However, extensive research on previous studies was carried out to extract as much data required for the assessment purpose as possible. The techniques applied in this study include sustainable assessment tools, GBI tools and in-depth review of literature in order to have a detailed understanding of the current situation of buildings in UTP.

4.0 Result and Discussion

This result was based on GBI Assessment Criteria for Non-Residential Existing Building (NREB) 1st Edition, January 2011, Version 1.1 which was conducted on Universiti Teknologi Petronas.

4.1 Energy Efficiency

An assessment of the energy efficiency is fully described in table 2. The total score for Energy Efficiency (EE) would be 3 out of 38. This is due to the fact that certain items cannot be evaluated as mentioned as shown in the table thereby affecting the outcome of the total score for Energy Efficiency (EE). Therefore, the marks for EE5, EE6, EE7 and EE9 will be excluded from the total score. The new total score for energy efficiency will be known as maximum possible score which is 10.

Table 2: Energy efficiency utilization

Energy Efficiency (EE)		
Item	**Justification**	**Source**
EE1	Evaluation not possible due to having no opportunity on BEIT software or other GBI approved software. UTP currently using Fuzzy control management.	Zazilah, Nor, Mohd, N., Jusoff, & Kamaruzaman. (2011). Optimal Operation of Chiller. *Recent Researches in Artificial Intelligence, Knowledge Engineering and Data Bases*, 109-115.
EE2	Lighting controls are control through energy management that are automatically turn on and off according to time.	Bhaskoro, P. T., Gilani, S. I., & Aris, M. S. (2013). Simulation of energy saving potential of a centralized HVAC system in an academic building . *Energy Conversion and Management*, 617-628.
EE3	No Sub-metering due to centralized energy control.	NA
EE4	Main energy source are from UTP GDC which do not use any renewable energy source.	Khamis, M. F., Baharudin, Z., Hamid, N. H., Abdullah, M. F., & Solahuddin, S. (2011). Electricity Forecasting For Small Scale Power System Using Artificial Neural Network . *The 5th International Power Engineering and Optimization Conference*, 54-59.
EE5	Evaluation not possible due to having no opportunity on BEIT software or other GBI approved software	NA

EE6	Required GBI recognized commissioning specialist (CxS) to perform the commissioning.	NA
EE7	Required GBI recognized commissioning specialist (CxS) to perform the commissioning.	NA
EE8	Currently using energy management system to log building system performance for HVAC system efficiency.	Khamis, M., Baharudin, Z., Hamid, N. H., Abdullah, M. F., & Nordin, F. T. (2011). Short Term Load Forecasting for Small Scale Power System Using Fuzzy Logic. *IEEE*, 1-5.
EE9	Non-available resource.	NA

4.2 *Indoor Environmental Quality*

Detail on indoor environmental quality is described in table 3. Hence the total score for Indoor Environmental Quality (EQ) would be 17 out of 21. This is due to deduction of points on low thermal comfort control, high glare and high noise level based on the IEQ and POE assessment.

Table 3: *Indoor Environmental Quality*

Indoor Environmental Quality (EQ)		
Item	**Justification**	**Source**
EQ1	Minimum indoor air quality (IAQ) performance meet the minimum requirements of ventilation rate and are contributing to the comfort and well-being of the occupants.	Khamidi, m. F., ismail, s. N., & mohd, n. (2013). Post occupancy evaluation (poe) and indoor environmental quality (ieq) assessment. *Journal design + built*, 1-21
EQ2	Prohibited smoking in all university area.	UTP management and policy.
EQ3	CO_2 monitoring had been done and meet minimum outside air requirement to maintain CO_2 level < 1,000 ppm	Khamidi, m. F., ismail, s. N., & mohd, n. (2013). Post occupancy evaluation (poe) and indoor environmental quality (ieq) assessment. *Journal design + built*, 1-21
EQ4	Using low VOC paints and coatings throughout the building and no added urea formaldehyde.	Kara, H. (2007). *University of Technology Petronas.* Bandar Seri Iskandar
EQ5	The building is fully naturally ventilated. Rainwater leakage through roof and walls.	Kara, H. (2007). *University of Technology Petronas.* Bandar Seri Iskandar
EQ6	No individual control on thermal comfort system.	Khamidi, M. F., Ismail, S. N., & Mohd, N. (2013). Post Occupancy Evaluation and Indoor Environmental Quality Assessment. *Journal Design + Built*, 1-21.
EQ7	The air change effectiveness meets the criteria. There are two types of mechanical ventilation system: variable volume constant	Khamidi, M. F., Ismail, S. N., & Mohd, N. (2013). Post Occupancy Evaluation and

	temperature air-conditioning, and 100 per cent fresh air constant volume variable temperature air handling.	Indoor Environmental Quality Assessment. *Journal Design + Built*, 1-21.
EQ8	>50% of the NLA has a daylight factor in the range of 1.o-3.5% as measured at the working plane.	Khamidi, M. F., Ismail, S. N., & Mohd, N. (2013). Post Occupancy Evaluation and Indoor Environmental Quality Assessment. *Journal Design + Built*, 1-21.
EQ9	94% of respondents claim that there is too much glare from artificial light. The assessments verify the result of POE.	Khamidi, M. F., Ismail, S. N., & Mohd, N. (2013). Post Occupancy Evaluation and Indoor Environmental Quality Assessment. *Journal Design + Built*, 1-21.
EQ10	According to MS1525, lighting for working interior for general offices is between 300-400 lux, while for class room 300-500 lux. However, the readings taken shows the illumination levels exceed the recommended value. This is most probably due to the exposure of the room towards the natural light (sunlight).	Khamidi, M. F., Ismail, S. N., & Mohd, N. (2013). Post Occupancy Evaluation and Indoor Environmental Quality Assessment. *Journal Design + Built*, 1-21.
EQ11	Different types of lighting are used: Laboratories: Surface-mounted fluorescent luminaries with low-brightness louvres Double-height Areas: High-bar luminaries Toilets /corridors: Low-voltage down lighters Library bookshelf: Linear fluorescent luminaries with low brightness louvers Lecture theatre: Low-voltage down lighters and wall-mounted floor washers Cafeterias: Modular fluorescent luminaries	Kara, H. (2007). *University of Technology Petronas.* Bandar Seri Iskandar
EQ12	>75% of the NLA has a direct line of sight through vision glazing at a height of 1.2m from floor level	Khamidi, M. F., Ismail, S. N., & Mohd, N. (2013). Post Occupancy Evaluation and Indoor Environmental Quality Assessment. *Journal Design + Built*, 1-21.
EQ13	For the lecture room, lab and lecturer room, the distraction comes from the blowers of air cons that produce high level of noise. The contribution of outdoor noise to indoor noise levels is usually small.	Khamidi, M. F., Ismail, S. N., & Mohd, N. (2013). Post Occupancy Evaluation and Indoor Environmental Quality Assessment. *Journal Design + Built*, 1-21.
EQ14	Reading for average temperature, relative humidity, CO 2 level and air velocity indicates that each element will contribute 1 point for GBI rating scores.	Khamidi, M. F., Ismail, S. N., & Mohd, N. (2013). Post Occupancy Evaluation and Indoor Environmental Quality Assessment. *Journal Design + Built*, 1-21.

EQ15	According to POE result, two parameters which are temperature and noise did not pass severity index boundary, means the most unsatisfied parameter.	Khamidi, M. F., Ismail, S. N., & Mohd, N. (2013). Post Occupancy Evaluation and Indoor Environmental Quality Assessment. *Journal Design + Built*, 1-21.

4.3 Sustainable Site Planning and Management

Based on these evaluated parameters, Sustainable Site Planning and Management (SM) score would be 7 out of 10 total score. UTP lost few marks due to lack acknowledgement on green vehicle priority and building user manual. Preferred car park should be provided for low-emitting and fuel-efficient vehicles. Also, UTP management needs to update building user manual as a guide and strategies for user information to sustain the building performance. This is described in table 4 below.

Table 4: Sustainable Site Planning and Management

Sustainable Site Planning and Management (SM)		
Item	Sustainable Site Planning and Management **Justification**	**Source**
SM1	The academic buildings have not been rated by any GBI or green rating system.	UTP
SM2	The building exterior are maintained by using environmentally nonpolluting chemical.	UTP
SM3	Erosion and sedimentation control for ongoing landscape operations including erosion and sedimentation are strictly monitored.	UTP
SM4	UTP does promote use of green vehicle however parking lots for green car is not provided.	UTP
SM5	UTP are providing parking for carpools and bus stop at every village, in front of Chancellor hall and main entrance.	Kara, H. (2007). *University of Technology Petronas.* Bandar Seri Iskandar
SM6	The roof design responds intelligently to the local climate by controlling the strong solar radiation and equally heavy downpours. Glazing is 10-millimeter single clear float glass with horizontal external shading to reduce solar transmission – a choice partly made possible by the protecting roof gesture.	Kara, H. (2007). *University of Technology Petronas.* Bandar Seri Iskandar
SM7	Currently, there is no documentation of green building design features and strategies for user information and guide to sustain performance during occupancy.	UTP management

4.4 Materials and Resources

The score for Materials and Resources (MR) is 5 out of 9 total score. However, MR1, MR2, MR3 parameters will be excluded. Therefore, it will be 5 out of 6 maximum possible scores. This is described in table 5.

Table 5: Materials and Resources

Items	Justification	Source
MR1	Non-available information	Na
MR2	Non-available information	Na

MR3	Non-available information	Kara, H. (2007). University technology Petronas. Badar Seri Iskandar
MR4		Na
MR5	Recycling bins are provided and UTP does not encourage minimization and recycling among occupants	UTP
MR6	The refrigerants are non-CFC or HCFC because the cooling system are from GDC chilled water.	Kara, H. (2007). University technology Petronas. Badar Seri Iskandar

4.5 Water Efficiency

Table 6 explains the water efficiency and availability. The total score for Water Efficiency (WE) is 6 from total of 12. However, WE4 and WE5 parameters with total score of 5 are excluded due to non-available data. Therefore, the score for Water Efficiency (WE) is 6 out of 7 maximum possible score.

Table 6: Water Efficiency

Water Efficiency (WE)		
Item	**Justification**	**Source**
WE1	Water is collected from the roof to be used for irrigation.	Kara, H. (2007). *University of Technology Petronas.* Bandar Seri Iskandar
WE2	Wastewater are being treated so that it can be recycled. Example are boiler water, cooling water and chilled water.	Kara, H. (2007). *University of Technology Petronas.* Bandar Seri Iskandar. Shahar, N. A. (2016, August 19). *GDC UTP*
WE3	Potable water consumption for irrigation are used from collected rainwater.	Kara, H. (2007). *University of Technology Petronas.* Bandar Seri Iskandar
WE4	No reference to utility calculations and water consumption record.	NA
WE5	No data available	NA

4.6 Innovation

Finally, for Innovation (IN), the score evaluated is 5 out of 10 total score. To summarizes, all the scores obtained is compared in the table to give a clearer view of current sustainability status of the academic building at Universiti Teknologi Petronas. To improve IN score, UTP should implement advanced air filtration technology, waterless urinals to be fitted to all male toilets. Central vacuum system that serves at least 50% of NLA also should be installed. Apart from that, fire system water during regular testing should be recycled and the use of self-cleaning facade and electro chromic glazed facade could help increase the innovation point.

Table 7: Innovation Introduced

Innovation (IN)		
Item	Justification	Source
IN1	UTP are currently using Co-generation system. Thermal mass storage system also accounting for at least 25% of total required capacity. Low ventilation system also available to reduce energy consumption.	Ambri, Z., & Yongo, W. (2011). Analytical Models for Energy Eudit of Cogeneration Plant. *Journal of Applied Sciences* Zahid, M., Ghazali, Z., & Rahman, H. U. (2017). A Step towards Sustainable University: A Case of Universiti Teknologi PETRONAS (UTP) Malaysia. *Global Business and Management Research: An International Journal*, 492-504
IN2	No engagement services of a Green Building Index in UTP at the time of this assessment.	UTP

5.0 Conclusion

Based on the results of GBI sustainable building assessment, Universiti Teknologi Petronas scores 43 out of 64 which is at 67% of total score hence the rating would be Silver. This result proves that some improvements need to be done on some parameters which did not passed the requirements. In a nutshell, Universiti Teknologi Petronas has high potential to meet GBI certified level that will help to increase the ranking status of university to higher level. This also would be a great opportunity for the university to set a benchmark as a role model of green university in Malaysia. However, this study recommends that UTP management develops corrective policies to improve the building overall performance and standard.

6.0 References

[1] Bragança L, Mateus R and Koukkari H 2010 Building Sustainability Assessment Sustainability 2010-2023
[2] Khamidi M F, Ismail S N and Mohd N 2013 Post Occupancy Evaluation and Indoor Environmental Quality Assessment Journal Design and Built 1-21
[3] Kara H 2007 University of Technology Petronas Bandar Seri Iskandar
[4] Ambri Z and Yongo W 2011 Analytical Models for Energy Eudit of Cogeneration Plant Journal of Applied Sciences
[5] Zazilah N, Mohd N and Kamaruzaman J 2011 Optimal Operation of Chiller. Recent Researches in Artificial Intelligence, Knowledge Engineering and Data Bases, 109-115
[6] Khamis M, Baharudin Z, Hamid N H, Abdullah M F and Nordin F T 2011 Short Term Load Forecasting for Small Scale Power System Using Fuzzy Logic IEEE 1-5
[7] Bhaskoro P T, Gilani S I. and Aris M S 2013 Simulation of energy saving potential of a centralized HVAC system in an academic building Energy Conversion and Management 617-628

[8] Bhaskoro P T, Gilani S I and Aris M S 2011 Electricity Forecasting for Small Scale Power System Using Artificial Neural Network The 5th International Power Engineering and Optimization Conference 54-59
[9] Shahar N A 2016 GDC UTP https://prezi.com/hx5rcaizg3gp/gdc-utp-by-nuraqilah/?webgl=0
[10] Okitsu J, Fatimie M, Zakaria N, Noano K and Abba A 2013 Towards an Architecture for Integrated Gas District Cooling with Data Center Control to Reduce CO_2 Emission IEEE
[11] Zahid M, Ghazali Z, and Rahman H U 2017 A Step towards Sustainable University: A Case of Universiti Teknologi PETRONAS (UTP) Malaysia Global Business and Management Research: An International Journal 492-504
[12] Yahya S S, Ariffin A and Ismail M 2009 Green Buildings in Campus: An Assessment of Green Potential for Existing Conventional Building Urban and Conservation Tropical Architecture Centre 1-17
[13] Long M 2016 Benefits of Green Building https://www.usgbc.org/articles/green-building-facts
[14] Administration U E 2017 U.S. Energy-Related Carbon Dioxide Emissions Washington: Independent Statistics and Analysis

Factors influencing customer intention to stay in green hotel in Malaysia

J M LIM[1], W Al Madhoun[2], C M Yee[1], G Nair[1], W S Siong[1] and H A Isiyaka[3]

[1] Department of Marketing, Faculty of Business and Finance, University Tunku Abdul Rahman, Malaysia
[2] Civil and Environmental Engineering Department, Universiti Teknologi Petronas, 32610 Seri Iskandar, Perak Darul Ridzuan Malaysia
[3] Department of Geography, Faculty of Earth and Environmental Sciences, Kano University of Science and Technology Wudil, Nigeria
***Corresponding author:** limds@utar.edu.my

Abstract. Customer stay intention (CSI) in green hotel needs a disclosure as there has been interest sought after by customer, society and business owners. Understanding customers drive and factors influencing their choices on green hotel is a very difficult task as it deals with complex human decision. No past studies had been done to assess customer stay intention (CSI) in Green Hotels in Malaysia as this study aims to provide factual findings for decision making and policy implementation. This study investigates the relationship between subjective norm (SN), perceived behavioral control (PBC), environmental awareness (EA), and perceived moral obligation (PMO) towards customer stay intention (CSI) in green hotel. A research model was developed to analyze the influence of SN,PBC,EA, and PMO towards CSI as extended from Theory of Planned Behavior (TPB). The survey was participated by 300 respondents that gathered what influenced CSI. These findings suggested that SN, PBC, EA, and PMO have positive influence towards CSI. Research deducted as environmental awareness (EA) has the strongest relationship with the customer stay intention. (CSI) in a green hotel and followed by the perceived moral obligation (PMO), perceived behavioural control (PBC) and subjective norms (SN) chronologically.

1. Introduction

With the changes of the era, the concerns of the society toward the environmental issues and disasters have progressively escalated [1]. Lately, Malaysia has been confronting tremendous amount of environmental issues like deforestation, climate change, and global warming. Malaysia's temperatures are predicted to increase by 0.6 C to 4.5 C by the year 2060 by the Intergovernmental Panel on Climate Change (2007) [2]. According to [3], global tourism sector has brought about the occurrences of harmful events that are eroding the health of the natural environment which are the global carbon emissions and the greenhouse emissions with the total percentage of 5% and 14% respectively. Thus, tourism is

Content from this work may be used under the terms of the Creative Commons Attribution 3.0 licence. Any further distribution of this work must maintain attribution to the author(s) and the title of the work, journal citation and DOI.
Published under licence by IOP Publishing Ltd

considered to have a major impact on the environment. In many countries, tourism is one of the major sources of income and employment. In recent years, numerous customers have been aware of the harming of environmental resources and environmental damages induced by the hotel industry [4]. A great alteration in the purchasing behaviors of customers and the customers' attitudes towards environmentally-friendly business establishment has occurred due to these environmental concerns and awareness [1]. Many customers had begun to change their purchasing habits by inquiring and purchasing for eco-friendly products because they have been realizing on the damages brought to the environment caused by their purchasing behavior [5].

Subsequently, hotels are turning up to be more concerned on keeping the balance among ecological problems, resources consumption problems, moral and societal issue and productivity. Mohamad et al (2014) [2] stated that the attention particularly in the sector of tourism towards the environment problems has result in the expanding number of the eco-friendly hotel. [6] characterized that Green hotel is a pro-environmental lodging property whose managers are keen to execute diverse green practices/programs that are water saving, energy saving and reducing the use of solid/ hazardous waste to assist in securing our planet and while at the same time to save cost. [7]. In recent years, numerous hotels around the world are putting intensive efforts in searching for compelling approaches to become "green" so as to position themselves distinctively within the competitive hotel industry and Malaysia is not an exception to this [8]. In the recent past, many counties have taken serious consideration relation to the environment and Malaysia was no exception. Malaysia had enacted the Environment Quality Act in the year 1974. Therefore, Malaysia has been considered as one of the earliest among those countries [9]. Likewise, in other countries, many hotels in Malaysia had incorporated the green concept in their business operations. There were several most known green hotels in Malaysia. The aim of this study is to understand the drive that affect customer's intention to stay in green hotels in Malaysia.

2.0 Methodology

2.1Data Collection and Analysis

The sampling strategy of the target group focus on respondents within the age eligible by law to book for hotel accommodation. The minimum age was taken as 18 years with no predefined maximum age. A well-structured questionnaire in the form of nominal, ordinal and Likert scale (section A, B and C) were designed and administered to the respondents in Klang valley and Ipoh. This study area was selected because of their high level of attraction and patronage. Non probability sampling technique was used to select the population size. As such, a total of 300 respondents were selected for the study. The data were analysed using descriptive statistics and inferential techniques (Pearson correlation and Multiple linear regression). The correlation analysis was devised by Pearson based on the value of correlation coefficient 'r'. The correlation between two parameters plotted on a XY scatter diagram can be termed as positive or negative. Correlation analysis is a common and useful statistical tool for assessing tourists' loyalty to the choice of hotel stay and comparing the relationships between certain critical variables [10]. The regression statistical technique is simply a measure to exhibit how well one variable predicts the other. It is use to ascertain to what extent an independent variable affects a dependent variable from coefficient of determination. It is use to form explicit equations that are less complex [11]. Using this method SN, PBC, EA, PMO were selected as the independent variable to understand their relationship and impact on CSI. Before then, a reliability test of the data was carried out to test if the data composition is fit for the analysis.

3.0 Result and Discussion

3.1 Demographic characteristics of the Respondents

In that sample size of 300 respondents, there were 157(52.33%) female A (47.67%) male respondents in total. 38.67% of survey participants from 31-40 years old had 116 respondents.102 respondents fall into income group RM 3001-RM4500.

3.2 Result for Pearson Correlation

Furthermore, the following correlation values are outcomes based on the analysis results: SN and CSI is r=0.57009 (p<0.0001); PBC and CSI is r=0.60253 (p<0.0001) and PMO and CSI is r=0.61292 (p<0.0001). In the correlation test, EA has found to be the most effect over CSI with the highest value of 0.64069. Due to the positive values shown in the results, all IVs are verified to have a positive influence towards the DV. Visitor loyalty to a unique place of choice remains an important indicator of successful destination development [13]

3.3 Result for Multiple Regression Model

Table 1 describe that F-vales is shown at the value of 117.68 with a significant table <0.001. All four constructs are shown to be significant which is SN, PBC, PMO, and EA. According to the table, EA has the strongest relationship towards CSI with the value of 0.33737. Hence, H1, H2, H3, and H4 are supported by the results.R^2 value formed in this analysis is 0.6147 which indicates that around 61% of the variation CSI in green hotel in Malaysia has been explain by the four variables. The following detailed observation can be deduced from the multiple regression model.

Table 1: Summary of result for multiple regression model.

Relationship	Parameter Estimated	Multiple Linear Regression	
	Results	Results	Remarks
H1: There is a significant relationship between SN and CSI in green hotel in Malaysia	0.22697	<.0001	Supported
H2: There is a significant relationship between PBC and CSI in green hotel in Malaysia	0.24261	<.0001	Supported
H3: There is a significant relationship between PMO and CSI in green hotel in Malaysia	0.28371	<.0001	Supported
H4: There is a significant relationship between EA and CSI in green hotel in Malaysia	0.33737	<.0001	Supported

3.3.1 H1: SN has a positive influence on the CSI in a green hotel in Malaysia

From table 1, results indicate that there is a significant relationship between SN and CSI. Han et al (2010) [4] further support this with their findings by defining that individuals tend to perceive social pressure from people who are considered significant to them. This is because of prominent referents are able to assist the individual in obtaining positive assessments of outcomes of the behavior for staying in an eco-friendly hotel. As also stated by Chang, (1998); Vallerand et al (1992); Ryu and Jang, (2006) [14,15,16], the positive or negative views of a person's important (referents) on the matter consider the performance of green behaviour would affect the formation of the favourable or unfavourable attitude toward staying in an eco-friendly hotel. Moreover, several past studies had found that a person's attitude towards an eco-friendly hotel and behavioural intention is significantly influenced by subjective norms [4,17,18,19]. Consequently, the results for H1 is supported as SN have a positive effect on CSI.

3.3.2 H2: PBC has a positive influence on the CSI in a green hotel in Malaysia

Referring to the outcomes from table 1, PBC is proven to have a significant influence towards CSI. This outcome is supported by [4,1,8] who discovered that the CSI is positively associated with his/her perceived control over the barriers for an eco-friendly hotel stay. This is also reinforced by other researchers such as Lien et al (2012) [20], Chen and Peng (2012) [21]. Moreover, Chen and Tung (2014) [5] states that one of the significant determinants that trigger CSI in a green hotel is the customer's perception of the ready availability of staying in a green hotel. In addition, the findings also justify that customers will have more intention to choose to stay in an eco-friendly hotel when it is considered easily accessible to them. Furthermore, Mohd (2015) [22] shared that a CSI in a green hotel is hinged on the accessible width of his/her time, resources, and opportunities to stay. In other words, customers control numerous factors, for instance, money, time and efforts that may influence his/her actual environmental behaviour and the staying intention in a green hotel. Untaru (2014) [23] also certify that the higher control on a particular behaviour of an individual due to the readiness of needed resources, his/her behavioural intention will be higher. Hence, H2 is supported as PBC has a positive effect on CSI.

3.3.3 H3: PMO has a positive influence on the CSI in a green hotel in Malaysia

PMO was shown to be statistically significant (P=<0.0001). As a result, H3 is supported. According to the findings of [24], moral obligation is considered as a vital element as the root for a pro-environmental behavioral disposition. Consumers would deem that they were morally obligated to secure the ecosystem and also prevent the used up of the limited natural resources. Prior researchers such as [25,26,27] has reported that moral obligation is an influential driving force of environmental behaviour. Thus, PMO had a positive impact on the CSI in an eco-friendly hotel. [5]. The results of the findings were also aligned with other studies such as [28, 29].

3.3.4 H4: EA has a positive influence on the CSI in a green hotel in Malaysia

The result shows a positive relationship between EA and CSI in a green hotel, whereby the result adheres to the findings of [17]. Han and Kim (2010) [17] indicate that environmental awareness is one of the variables to ameliorate the attitude towards the green business establishment. Thus, the higher the customer's awareness towards the environment, the higher the staying intention in green hotels will be. Besides, there were several past researchers [30,31,32,33,34,35] who had studied on the green marketing, green consumer behaviour and environmental psychology that have asserted the significance of the consumer's environmental awareness, environmental behaviours, perceived effectiveness and environmentally-friendly reputation during the occasion of making decision to involve in an environmentally responsible buying behaviour. Furthermore, the results from the statistical analysis are also supported by [5]. As mentioned by the researchers, the attitude of customers to visit an eco-friendly hotel positively depends on their personal concern towards the environment. In addition, a favourable attitude toward visiting the green hotel was formed the minute environmental concern is of high level. Therefore, H4 is supported and have been demonstrated that there is a positive impact on the CSI in a green hotel by EA.

4.0 Conclusion

By conducting this study, an insight about the factors that affect CSI in Malaysia's green hotel is provided. The result show that SN, PBC, PMO, and EA have a significant correlation with the CSI.

5.0 References

[1] Han H and Kim Y 2010 An investigation of green hotel customers' decision formation: Developing an extended model of the theory of planned behavior *International Journal of Hospitality Management 29* 659–668

[2] Mohamad Z Z, Arifin T R T, Samsuri A S and Munir M F M B 2014 Intention to Visit Green Hotel in Malaysia: The Impact of Personal Traits and Marketing Strategy *International Journal of Business and Social Science 5*(7) 157-173

[3] Lindqvist J and Andersson M 2015 Travelling green: Variables influencing students' intention to select a green hotel (Dissertation) Retrieved from http://urn.kb.se/resolveurn=urn:nbn:se:hkr:diva-14980

[4] Han H, Hsu L and Sheu C 2010 Application of the Theory of Planned Behaviour to green hotel choice: Testing the effect of environmental friendly activities. *Tourism Management* 31(3) pp.325-334

[5] Chen M F and Tung P J 2014 Developing an extended Theory of Planned Behaviour model to predict consumers' intention to visit green hotels. *International Journal of Hospitality Management*, 36, pp. 221-230

[6] Green Hotels Association 2014 What Are Green Hotels Retrieved from http://www.greenhotels.com/whatare.htm

[7] Rahman I and Reynolds D 2016 Predicting green hotel behavioural intentions using a theory of environmental commitment and sacrifice for the environment *International Journal of Hospitality Management* 52 107-116. doi: 10.1016/j.ijhm.2015.09.007

[8] Eirini T and Antonia D 2014 "Do beliefs affect customers' intentions to choose green hotels?" *International Journal on Strategic Innovative Marketing* 1 51-60.

[9] Punitha S, and Rahman A 2011 Antecedents of Green Purchasing Behaviour among Malaysian Consumers *International Business Management 5*(3) 129–139

[10] Prayag G, and Ryan C 2012 Antecedents of Tourists' Loyalty to Mauritius: The Role and Influence of Destination Image Place Attachment Personal Involvement and Satisfaction Journal of Travel Research 51(3) 342–356 DOI: 10.1177/0047287511410321

[11] Isiyaka H A, Juahir H, Toriman M E, Azid A, Gasim B M and Kamarudin M K A 2015 Assessment of the Spatial Variation and Source Apportionment of Air Pollution Based on Chemometric Techniques: A Case Study in the Peninsular Malaysia Jurnal Teknologi 77:1 (2015) 33–44 | www.jurnalteknologi.utm.my

[12] Mustaffa N I H, Latif M T, Ali M M and Khan M F 2014 Source Apportionment of Surfactants in Marine Aerosols at Different Locations Along the Malacca Straits Environmental Science and Pollution Research 21(10): 6590-6602

[13] Poira Y, Reichel A and Biran A 2006 "Heritage Site Perceptions and Motivations to Visit" Journal of Travel Research 44: 318-26

[14] Chang M K 1998 Predicting unethical behaviour: A comparison of the theory of reasoned action and the theory of planned behavior *Journal of Business Ethics 17*: 1825-1834.

[15] Vallerand R J, Deshaies P, Cuerrier J P., Pelletier L G and Mongeau C 1992 Ajzen and Fishbein's theory of reasoned action as applied to moral behaviour: A confirmatory analysis Journal of Personality and Social Psychology 62(1) 98-109

[16] Ryu K and Jang S 2006 Intention to Experience Local Cuisine in a Travel Destination: The Modified Theory of Reasoned Action. *Journal of Hospitality & Tourism Research 30*(4) 507-516 doi: 10.1177/1096348006287163

[17] Han H and Kim Y 2010 An investigation of green hotel customer's decision formation: Developing an extended model of the theory of planned behavior *International Journal of Hospitality Management, 29*, 659–668

[18] Wu K S and Teng Y M 2011 Applying the Extended Theory of Planned Behavior to Predict the Intention of Visiting a Green Hotel African *Journal of Business Management* 5 17 pp. 7579-7587

[19] Teng YM, Wu K S and Liu H H 2013 Integrating altruism and the theory of planned behavior to predict patronage intention of a green hotel Journal *of Hospitality & Tourism Research* Vol 37 No 1 pp 1-18

[20] Lien C, Huang C and Chang H 2012 The influence of green consumption cognition of consumer on behavioural intention – a case study of the restaurant service industry *African Journal of Business Management* Vol 6 No 26 pp 7888-7895

[21] Chen A and Peng N 2012 Green hotel knowledge and tourists staying behaviour *Annals of Tourism Research* Vol 39 No 4 pp 2211-2216

[22] Mohd Suki 2015. Consumers environmental behaviour towards staying at a green hotel *Management of Environmental Quality: An International Journal 26* (1) 103-117 doi: 10.1108/meq-02-2014-0023

[23] Untaru E N, Epuran G and Ispas A 2014 A conceptual framework of consumer's pro-environmental attitudes and behaviors in the tourism context Bulletin of the Transylvania University of Braşov *Series V Economic Sciences 7*(2) 85-94

[24] Tan BC and Lau TC 2010 Attitude towards the Environment and Green Products: Consumers Perspective *Manage Sci Eng 4*(2): 27-39

[25] Hopper J R and Nielsen J.M 1991 Recycling as altruistic behavior: Normative and behavioral strategies to expand participation in a community recycling program. *Environment and Behavior 23* 195-220

[26] Vining J and Ebreo A 1992 Predicting recycling behavior from global and specific environmental attitudes and changes in recycling opportunities *Journal of Applied Social Psychology* 22 1580-1607

[27] Stern P and Dietz T 1994 The Value Basis of Environmental Concern *Journal of Social Issues 50*(3) 65-84 doi: 10.1111/j.1540-4560. 1994.tb02420x

[28] Haines R, Street M D and Haines D 2008 The influence of perceived importance of an ethical issue on moral judgment moral obligation and moral intent *Journal of Business Ethics 81* (2) 387–399.

[29] Lam S P 1999 Predicting intentions to conserve water from the theory of planned behavior perceived moral obligation and perceived water right Journal *of Applied Social Psychology 29* (5) 1058–1071

[30] Schlegelmilch B B, Bohlen G M and Diamantopoulos A 1996 The link between green purchasing decisions and measures of environmental consciousness *European Journal of Marketing 30*(5), 35-55

[31] Roberts J A 1996 Green consumers in the 1990 Profile and implications for advertising *Journal of Business Research 36* 217–231

[32] Takacs S A 2007 Barriers to environmental concern *Research in Human Ecology 14*(1) 26–38

[33] Ryan A and Spash C L 2008 Measuring awareness of environmental consequences Two scales and two interpretations Socio-economics and environment in discussion (SEED) working papers Canberra Australia Commonwealth Scientific and Industrial Research Organisation (CSIRO)

[34] Paco A F, Raposo M L and Filho W L 2009 Identifying the Green Consumer A Segmentation Study *Journal of Targeting Measurement an* diately following the last numbered section of the paper.

Current and potential impacts of sea level rise in the coastal areas of Malaysia

Sofia Ehsan*[1], Rawshan Ara Begum[2], Nor Ghani Md Nor[3] and Khairul Nizam Abdul Maulud[1,4]

[1]Institute of Climate Change, Universiti Kebangsaan Malaysia, Bangi, Selangor, Malaysia
[2]Center for Water Cycle, Marine Environment and Disaster Management (CWMD), Kumamoto University, 2-39-1 Kurokami, Chuo-ku, Kumamoto 860-8555, Japan
[3]Chancellory, Universiti Kebangsaan Malaysia, Bangi, Selangor, Malaysia
[4]Faculty of Engineering and Build Environment, Universiti Kebangsaan Malaysia, Bangi, Selangor, Malaysia

*corresponding author, E-mail: sofia.ehsan@outlook.com

Abstract. Sea level rise is one of the most concerning and costly effects of climate change that impacts the sustainable development of coastal areas. Malaysia, representing 13% of the total land area within 5 km of a coast, is threatened by the devastating impacts of sea level rise. This study attempts to highlight the current and potential impacts of sea level rise in several high risk coastal areas of Malaysia. Currently, coastal erosion and coastal flooding are the major effects of sea level rise impacting the important coastal infrastructure. The coast of Selangor and Batu Pahat experienced severe coastal erosion recording the total eroded area of 1878.5 hectares and 415.47 hectares respectively. Likewise, the coastal flooding in Johor coastal flood was damaging an estimated RM 0.35 billion worth of infrastructure and RM 2.4 billion of economic losses. In addition, one meter rise in sea level is expected to cause the loss of 180,000-hectare of agricultural land, 15%-20% of mangrove forests loss along the coastline. Among the selected high risk areas along the coast of Malaysia, Batu Pahat is estimated to experience 100% loss of the development area followed by Port Klang (40.67%), Kedah (38.57%), Kuala Terengganu (4.86%), Kota Kinabalu (4.46%) and Kuching (2.64%). The potential sea level rise will amplify the existing impacts and create new risks for coastal population and development. Thus, proper adaptation measures are necessary in order to reduce the adverse impacts and economic costs of sea level rise in Malaysia.

1. Introduction

The global climate is changing at rates unprecedented in recent human history [1]. Sea level rise is one of the most concerning and costly effects of climate change, which has four major impacts in coastal areas: coastal flooding, coastal inundation, coastal erosion, and saltwater intrusion. While these impacts already exist in coastal regions, sea level rise intensifies their severity [2]. It addition, it will cause changes in the intensity and frequency of extreme events from combined effects of high spring tides, storm surges, surface waves and river flooding [3].

Content from this work may be used under the terms of the Creative Commons Attribution 3.0 licence. Any further distribution of this work must maintain attribution to the author(s) and the title of the work, journal citation and DOI.
Published under licence by IOP Publishing Ltd

These pose major threat to the sustainable development of coastal areas as they directly impact the physical property, important coastal infrastructure including transport network and utility infrastructure, production process, natural and cultural resources, and indirectly, impacts the ecosystem, economic activities, income, wealth, public health, safety, and overall social well-being of the coastal community. For instance, sea level rise directly affects the design and structural stability of the existing infrastructure such as coastal roads, revetment, groin, breakwater, coastal tide gates, ports, fishing pier, beach and also commercial and residential buildings [4] in several ways such as accelerated degradation and greater likelihood of cascading failures (e.g. failures or destruction of critical infrastructure, immobilization due to transportation system breakdown, blackouts due to power grid failures, and catastrophic saltwater contamination of water supplies) [5]. Below figure 1 illustrates the climate change process and major impacts of sea level rise in the coastal areas.

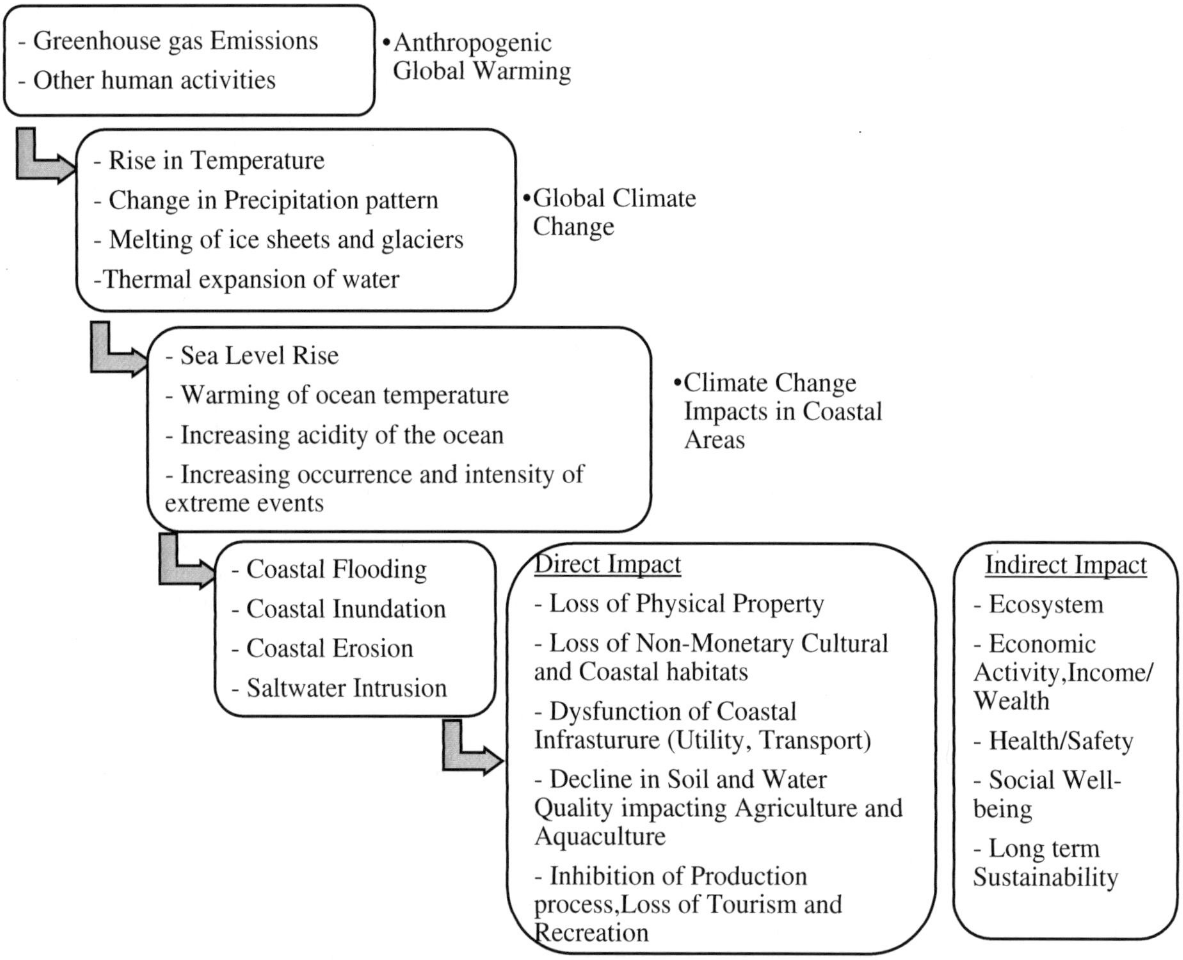

Figure 1. Sea level rise and its impacts in the coastal areas

The rising sea levels phenomena have significant long term impacts on the coastal area and their severity depends on the extreme, exposure and vulnerability of coastal areas [6]. Low-lying coastal areas are particularly susceptible to storm surge and flooding, and the effects on communities can be catastrophic. In the future, such effects will be intensified by rising sea levels [7] and may lead to prolonged disruption of economic activities, livelihoods, and long term sustainability of the coastal communities. Sea level rise threatens a significant number of people and assets in coastal areas. This study will highlight the sea level rise trend and the current and potential impacts of sea level rise in the coastal areas of Malaysia.

2. Coastal areas of Malaysia

The coastal zone in Malaysia is broadly defined as the areas where terrestrial and marine processes interact such as coastal plains, deltaic areas, coastal wetlands, estuaries and lagoons. Malaysia, with a land area of 330,000 km^2 and a coastline of 4800 km and its coastal areas occupies a total land area of 4.4 million hectares representing 13% of the total land area within 5 km of a coast, approximately 1.2 million hectares of the coastal areas are found in Peninsular Malaysia, 1 million hectares in Sabah and 2.2 million hectares in Sarawak. About 70% of the total population live in the coastal zones. There are 22 urban settlements and 12,400 rural settlements, 28 industrial estates and 54 ports located in the coastal zone of Peninsular Malaysia. Moreover, the development along the coasts and surrounding areas has been extensive such as agricultural development (especially rubber, coconut, paddy, and oil palm plantation), fisheries and aquaculture, manufacturing industries, transportation and communication, oil and gas reserves, mineral resources e.g. tin, gold, platinum, chromite etc, natural habitat and tourism resources [8].

3. Sea level rise (SLR) trend in Malaysia

Sea level has shown a rising trend that has been observed globally and along Malaysian coastline. For instance, according to Intergovernmental Panel on Climate Change (IPCC) [1], globally sea level has risen by 2.4 mm to 3.8 mm (per year). On the other hand, the study about the impact of sea level rise in Malaysia on year 2010 by National Hydraulic Research Institute of Malaysia (NAHRIM) reported the average rise in sea level along 30 stations in Malaysia from 1993 to 2010 as 2.73 mm to 7.0 mm (per year) and found that there has been an important increase in SLR trends over the last 5 years compared to SLR trends over 20 years ago [9]. Table 1 shows the observed and projected SLR Trend around the global and in Malaysia.

Table 1. The observed and projected SLR Trend (Global vs Malaysia)

Sea Level Rise (SLR) Rate (mm/year)	Global IPCC (AR4 2007)	Malaysia NAHRIM (2010)	
Observed	2.4 – 3.8*	2.73 – 7.0* (1993-2010)	
		West	East
Projected by 2100	2.6 – 5.9	2.5 – 5.2	4.3 – 10.6
*measured by Satellite Altimeter			

In addition, IPCC [10] has recently revised the potential SLR and estimated that global sea level rise may range from 0.3 m to approximately 1 m by 2100 based on a better understanding of glacial and ice sheet input. Consequently, climate change caused by the enhanced greenhouse impact will most likely accelerate the rise in sea levels in the future.

Similarly, NAHRIM predicted the rates of future SLR for 30 stations along Malaysia's coast using linear analysis and observation data from altimetry satellite (figure 2) that shows the SLR projection for the year 2100 which is estimated to be 0.25 - 0.5 m with maximum value occurring in low-lying areas along the Northeast coast (Kelantan State) and West coast (Kedah State) of Peninsular Malaysia. The maximum value of SLR in Sabah on year 2100 is estimated as 0.690 – 1.06 m occurs at Tawau, Semporna, Lahad Datu, Sandakan and Kudat as known as low-lying areas, river mouths and estuaries of the East coast; and in Sarawak, the maximum value of the SLR projection on year 2100 is 0.43 - 0.64 m occurs at Southwest coast consists of Meradong, Batang Iagan and Batang Rajang [9].

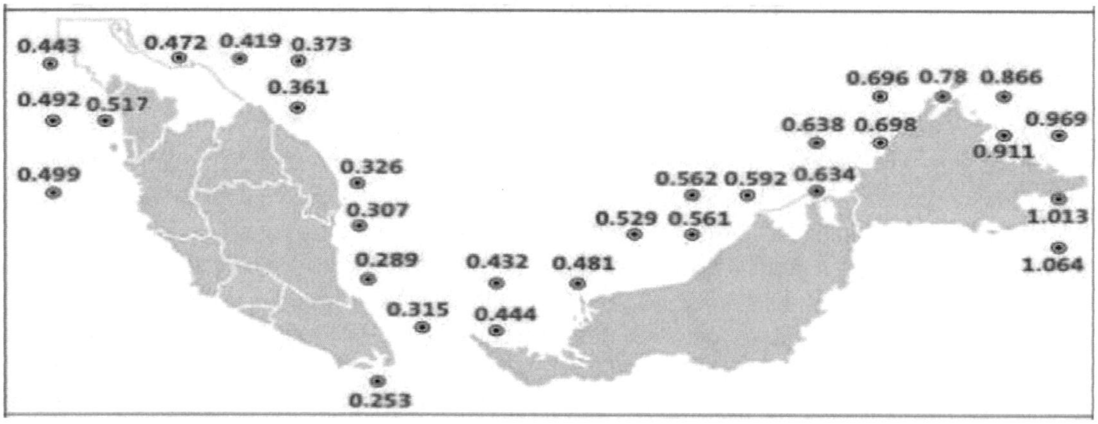

Figure 2. Projected sea level rise for Malaysia coastlines by year 2100 (Adapted From [9])

4. Current impacts of SLR in Malaysia

Coastal land is a valuable resource that has important economic, social, cultural and ecological values. However, several impacts sea level rise are already being experienced by the coastal communities of Malaysia mostly due to coastal erosion and coastal flooding putting millions of lives and billions of dollars' worth of property and infrastructure at risk.

4.1 Coastal erosion

Selangor, one of the major states on the west coast of Peninsular Malaysia is the largest economy in Malaysia in terms of gross domestic product (GDP) with RM 239.968 billion in 2015, making up 22.6% of the total GDP of Malaysia [11]. This state is also the most developed in Malaysia with good infrastructure such as highways and transport, has the largest population with a high standard of living and the lowest poverty rate in the country [12]. However, the coast of Selangor measuring 276 km is threatened by the climate change impact. For instance, the shoreline has being changed during 1984-2013 recording the total eroded area of 1878.5 hectares and total accretion area of 2447.4 hectares. The coastal area of Kapar, Kelang, Jugra, Telok Panglima Garang and Kelanag in Kuala Selangor and Klang districts are among the highest eroded areas [13]. At the coast of Selangor, the monitoring carried out proves that strong ocean currents, waves and winds damages the infrastructure near the coastal area. Figure 3 illustrates the effects of coastal erosion at Bagan Pasir, Selangor.

Figure 3. Impact of coastal erosion to the residential property at Bagan Pasir, Selangor (Adapted from [13])

In addition, Johor, the third biggest state economy in Malaysia with GDP of RM104.4 billion, which accounted for 9.4% of Malaysia's GDP is also threatened by the adverse impacts of sea level rise. Batu Pahat is one of the small districts of Johor state located at the south of Malaysia. A study was conducted by Khairul Nizam and Roshana (2015) found that Batu Pahat experienced severe coastal erosion. This area experienced 415.47 hectares of erosion and 68.52 hectares sedimentation over 3 years from 2011 to 2013. Based on their research, it is clear that 85.84 percent of Batu Pahat is extremely experiencing erosion phenomena occurred only within 3 years [14]. Figure 4 shows the phenomenon of coastal erosion and its impacts to the infrastructure damage occurring in Batu Pahat coastal areas.

Figure 4. Coastal erosion and its impact to the coastal infrastructures at Batu Pahat, Johor (Adapted from [14])

4.2 Coastal flooding

Impacts of sea level rise to the coastal communities due to coastal flooding has also become a serious issue in Malaysia. Floods are the most disastrous natural hazard in Malaysia. Even though, most floods occurred in Malaysia as a result of cyclical monsoons bringing heavy and regular rainfall and inadequate drainage system to curb with the excessive rainfall intensifying flooding problem [15], a series of flood seen in Johor state of Malaysia in 2006-2007 are believed to be extraordinary events that are closely related to global climate change and sea level rise impacts [16]. The coastal flooding in Johor during 2006-2007 with the worst case in Kota Tinggi and Segamat was caused due to the heavy rainfall within a long period of time, which occurred in the low area and the condition of the shallow river and high tide impacts from sea level rise [17]. Figure 5 illustrates coastal flooding and its impact to the coastal infrastructures.

Figure 5. Coastal flooding and its impact to the coastal infrastructures at Johor 2006-07 (Adapted from [18])

The impacts of the coastal flood was devastating and affected more than 100,000 victims in two separate events [18], damaging an estimate of RM 0.35 billion worth of infrastructure and RM 2.4 billion of economic losses [19]. In another study, the damage cost of 2006- 2007 Johor flood was estimated at RM1.5 billion [20]. In addition, an estimate of RM84 million of agriculture produce were damaged or losses affecting 7000 farmers. The loss of agricultural production due to inundation of erosion during the floods, in Johor signified RM 46 million for western Johor Agricultural Development Project Area [21]. Moreover, estimated cost to repair the transport infrastructure (road and bridges) was RM147 million by Public Work Department (PWD) and hydraulic structures was RM260 million by Department of drainage and Irrigation (DID) [18].

In addition, another climate change impacts related to sea, which causes coastal flooding in low lying coastal areas is known as La Nina phenomena. According to the Malaysian Meteorological Department, La Nina is caused by cooling of Pacific Ocean and causing higher rainfall compared to normal weather [22]. In 2011 and 2012, coastal flooding due to La Nina phenomena caused floods in Johor -the four worst affected areas were Segamat, Batu Pahat, Kluang and Muar [23]. One of the major impacts of this flooding was the effect to the palm oil production. Ayat K et al. (2012) estimated the potential income losses to the palm oil production at RM155.10 million and RM168.22 million during the two years [24]. In addition, flood affected estates had spent RM25.80 million and 26.48 million respectively to repair the infield roads.

Moreover, coastal flooding as a result of high tides meeting high water flow is prevalent at coastal cities such as Klang, Teluk Intan and several places in Penang [15]. For instance, 'king tides', the higher than normal waves that threaten low lying areas damaged coastlines in Penang, Kedah, Perak, Selangor, Negeri Sembilan, Melaka, Johor in 2016. Many of these coastal areas are densely populated and rich coastal infrastructures such as Penang is an area with the most intense coastal population (e.g. 860 people per square kilometer) [25]. However, coastal flooding severely impacts many aspects of these low lying coastal areas. Therefore, sea level rise may reduce the size of these coastal areas and its' infrastructure; worsen inundation, erosion, and other coastal hazards; threaten vital infrastructure, and thus compromise the socio-economic wellbeing of the island communities [26].

5. Potential impacts of SLR in Malaysia

Numerous areas in Peninsular Malaysia, especially coastal areas are at risk of being eroded, flooded and inundated if the ice coating in Greenland is fully melted. Several studies estimated the likely impact of sea level rise in Malaysia. For instance, Nicholls and Mimura (1998) has concluded that one meter increase in sea level rise will cause 700 square kilometer (2.1%) of land loss and displaces more than 0.3% of the population in Malaysia [27].

Moreover, one meter rise in sea level will also cause the loss of 180,000-hectare of agricultural land, 15%-20% of mangrove forests loss along the coastline, and possible relocation of shore-based power stations [28]. Besides, there are studies that focuses on the potential sea level rise impacts to the most vulnerable coastal areas in Malaysia. For example, according to the National Coastal Vulnerability Index Study [29], an average of 30% land loss is estimated at Langkawi Island and Tangjung Pia in Malaysia based on one meter sea level rise scenario.

In addition, based on the SLR projections NAHRIM carried out desktop studies on selected high risk areas along the coast of Malaysia e.g. mud-flat area at Batu Pahat (Sg. Senggarang to Sg. Punggur), active development and high population area at Kelang, Selangor, Kedah Estuary, Terengganu Estuary, Kota Kinabalu, and Kuching [30]. Utilizing numerical modelling suite and GIS software, potential SLR inundation maps were produced to assess the SLR towards population, development, mangrove area and paddy field that exist within these areas. Table 2 describes the potential impact of SLR in vulnerable areas in Malaysia.

With the potential 1 meter increase in sea level rise, Batu Pahat is estimated to experience 100% loss of the development area followed by Port Klang (40.67%), Kedah (38.57%), Kuala Terengganu (4.86%), Kota Kinabalu (4.46%) and Kuching (2.64%). In addition, a significant number of population are threatened to be inundated due to 1 meter sea level rise, e.g. 22000 in Port Klang, 12000 in Kedah, 5735 in Batu Pahat, 5000 in Kuala Terengganu, 2000 in Kuching, 1000 in Kota kinabalu are at risk due to future sea level rise. Similarly, all the mangrove areas are also likely to be inundated in Batu Pahat and Kedah.

Table 2. Potential impact of SLR in vulnerable areas (adapted from [30])

Study Area (Km^2)		Impact of Sea Level Rise (Inundation)			
		Development Area (Km^2)	Population	Mangrove Area (Km^2)	Paddy (Km^2)
Batu Pahat (Existing)	46	6.28	5,735	0.8	
0.253m SLR[1]	9.2 (20%)	0.884 (14.08%)	600 (10.46%)	0.716 (89.87%)	
0.5m SLR	22 (47.83%)	3.933 (62.62%)	3,080 (53.7%)	0.8 (100%)	
1m SLR[2]	46 (100%)	6.28 (100%)	5,735 (100%)	0.8 (100%)	
Port Klang (Existing)	360	107	49,000	161	
0.5m SLR[1]	17.14 (4.7%)	2.28(2.13%)	4,000(8.16%)	14.27(8.8%)	
1m SLR[2]	87.6 (24.4%)	14.36 (13.4%)	13,000 (26.5%)	37.56 (23.8%)	
2.2m SLR[3]	146.42 (40.67%)	31(29%)	22,000 (44.9%)	116.97 (72.8%)	
Kedah (Existing)	70	28.77	22,000	0.6	47.84
0.5m SLR	12.51 (17.87%)	2.53 (8.79%)	4,000 (18.18%)	0.6 (100%)	9.38 (19.61%)
1m SLR	27 (38.57%)	7.06 (24.54%)	12,000(54.55%)	0.6 (100%)	16.78 (35.08%)
Kuala Terengganu (Existing)	147	36.87%	36,000		
0.5m SLR	3.46 (2.35%)	3.26 (8.84%)	4,000 (11.11%)		
1m SLR	7.15 (4.86%)	6.81 (18.47%)	5,000 (13.89%)		
Kota Kinabalu (Existing)	32.71	17.3	9,000		
0.5m SLR	0.36 (1.1%)	0.09 (0.52%)	<1,000 (<11.11%)		
1m SLR	1.46 (4.46%)	0.41 (2.37%)	1,000 (11.11%)		
Kuching (Existing)	576	72.57	88,000		
0.5m SLR	14.09 (2.45%)	4.07 (5.61%)	1,000 (1.14%)		
1m SLR	15.2 (2.64%)	4.53 (6.24%)	2000 (2.27%)		

Note: 1 refers to the mean SLR in year 2100, 2 refers to the sensitivity SLR, 3 refers to the max SLR in year 2100.

6. Adaptation measures to reduce the impacts of SLR

The potential sea level rise impacts reflect that rising sea level in Malaysia and around the world will amplify these existing impacts and create new risks for coastal population and development. Therefore, the economic cost of sea level rise along the coast of Malaysia without any adaptation measures could be huge as the previous studies found out in the other coastal areas of the world. For instance, the

IOP Conf. Series: Earth and Environmental Science **228** (2019) 012023 doi:10.1088/1755-1315/228/1/012023

economic cost of US$20.4 billion for the coast of United States [31], US$23 million for the coast of Montevideo, Uruguay [32], and US$30 billion for the total cost of land loss for the coasts in Poland [33]. MOSTE (2000) estimated the economic costs from socio-economic impacts based on the high rate of sea level rise (0.9 cm/yr) in Malaysia which projected the agricultural production loss at RM46 million from eroded/inundated lands in Western Johor Agricultural Development Project area, which accounts for about 25% of the national drainage areas. In addition, long-term annual flood damage estimated at about RM88 million in Peninsular Malaysia and RM12 million for Sabah/Sarawak based on 1980 price level, which will be increased by 1.67 times if the flood frequency is doubles. Similarly, loss of fisheries production is estimated at RM300 million based on 2% loss of mangrove resulting in a loss of about 70,000 tonnes of prawn production values at RM4,500/ tonne [28]. Therefore, in order to offset the damage due to climate change, adaptation strategies are necessary measures which refers to any adjustments to human activities that reduce the vulnerability of humans and ecosystems to climate change impacts [1].

Over the last decades, adaptation has become one of the central approaches to reduce the adverse impact of changing global climate, has had a widespread benefit in reducing society's vulnerability to coastal hazards [34] and enables coastal communities to reduce its detrimental impacts by averting or reducing the potentially negative consequences, while benefiting for potentially positive consequences [35]. There are several coastal adaptation measures such as protection approaches to defend vulnerable areas using seawalls, sea dikes, storm surge barriers and closure dams and land claim etc.; accommodation approaches to occupy vulnerable areas but accepting the greater degree of flooding by changing land use, construction methods and improving preparedness such as flood-proofing, wetland restoration, floating agricultural systems, flood hazard mapping, flood warnings etc.; and retreat approaches to abandon currently developed areas, resettle inhabitants such as managed realignment and coastal setbacks. Figure 6 lustrates adaptation strategies in coastal areas [36].

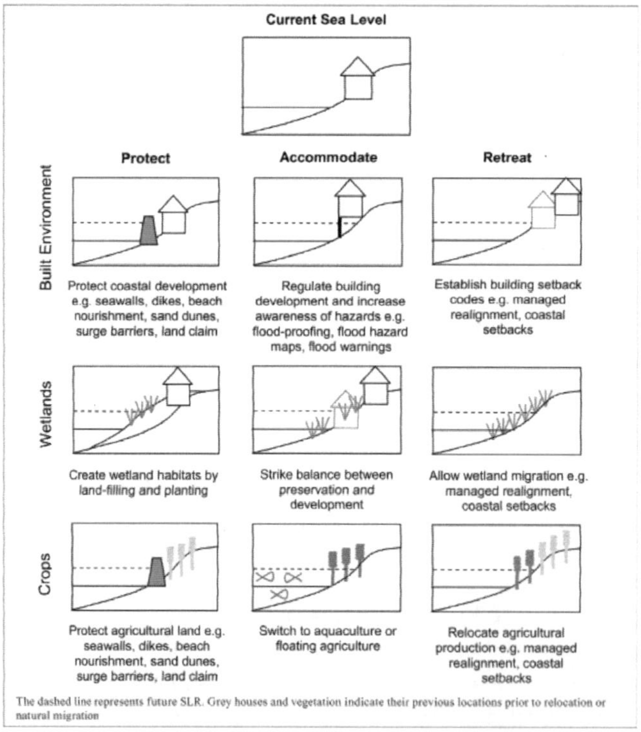

Figure 6. Coastal adaptation measures (Source: Adapted from [36])

7. Concluding Remarks

Climate change in coastal areas such as sea level rise, increased occurrence and intensity of storms and high tide poses major concern as they result in coastal erosion, flooding, salt water intrusion and inundation which in turn impacts the important infrastructure, physical property, production processes and wetland services. The economic cost of current and potential SLR in Malaysia is enormous. Nevertheless, noted in Malaysia's Second National Communication (NC2) to the United Nations Framework Convention on Climate Change (UNFCCC), there is renewed interest in climate change adaptation needs for Malaysia [37]. Similarly, the Malaysian Second National Communication and Fourth Assessment Report of Intergovernmental Panel on Climate Change (IPCC) stressed on research gaps and needs in terms of socio-economic impacts and response as well as comprehensive estimation of adaptation costs and benefits [38] due to its importance in decision-making on economic viable adaptation option, future investment decision, fund disbursement and negotiations at the global level. Therefore, cost-efficient and location specific appropriate adaptation measures for the coastal community are necessary to reduce the cost of climate change among the prevalent adaption measures such as sea wall, living shoreline, road elevation, house elevation, investment avoidance and public purchases.

Acknowledgements
The research was funded by Ministry of Higher Education Malaysia through its projects Transdisciplinary Research Grant Scheme TRGS/1/2015/UKM/02/5/3.

Reference

References

[1] IPCC 2007 *Climate change 2007: impacts, adaptation and vulnerability. contribution of working group ii to the fourth assessment report of the Intergovernmental Panel on Climate Change,* ed M L Parry *et al* (Cambridge: Cambridge University Press) p 976

[2] Almeida B A and Mostafavi A 2016 Resilience of infrastructure systems to sea-level rise in coastal areas: impacts, adaptation measures, and implementation challenges *Sustainability* **8(11)** 1115

[3] Church J A, Clark P U, Cazenave A, Gregory J M, Jevrejeva S, Levermann A, Merrifield M A, Milne G A, Nerem R S, Nunn P D, Payne A J, Pfeffer W T, Stammer D and Unnikrishnan A S 2013 *Sea Level Change In Climate Change 2013: The Physical Science Basis. Contribution of Working Group I to the Fifth Assessment Report of the Intergovernmental Panel on Climate Change,* ed T F Stocker *et al* (Cambridge: Cambridge University Press)

[4] Nor Aslinda Awang and Elyasabath Mohd Ghazali 2013 Monitoring & evaluation of climate change adaptation initiative for Malaysia

[5] Abel N, Gorddard R, Harman B, Leitch A, Langridge J, Ryan A and Heyenga S 2011 Sea level rise, coastal development and planned retreat: Analytical framework, governance principles and an Australian case study Environ. Sci. Policy **14** 279–88

Tacoli C 2009 Crisis or adaptation? Migration and climate change in a context of high mobility *Environ. Urban.* **21** 513–525

Tol RSJ 2009 The economic effects of climate change J. Econ. Perspect. **23** 29–51

Gornitz V, Couch S, Hartig E K 2001 Impacts of sea level rise in the New York City metropolitan area Glob. Planet. Chang. **32** 61–88

[6] Faour, Ghaleb, Fayad, Abbas, Mhawej and Mario 2013 GIS-Based approach to the assessment of coastal vulnerability to sea level rise: case study on the Eastern Mediterranean *Journal of Surveying and Mapping Engineering* **1(3)** 41-48

McLean R F, Tsyban A, Burkett V, Codignotto J O, Forbes D L, Mimura N, Beamish R J and

Ittekkot V 2001 *Coastal zones and marine ecosystems.* in *climate change 2001: impacts, adaptation, and vulnerability, contribution of working group ii to the third assessment report of the IPCC,* ed J J McCarthy *et al* (Cambridge: Cambridge University Press) p379

IPCC 2012 Managing the risks of extreme events and disasters to advance climate change adaptation. a special report of working groups i and ii of the Intergovernmental Panel on Climate Change, ed C B Field et al (Cambridge: Cambridge University Press) p582

[7] Nicholls R J, Hoozemans F M J, Marchand M 1999 Increasing flood risk and wetland losses due to global sea level rise: regional and global analyses *Glob. Environ. Change* **9** 69- 87

Nicholls R J, Cazenave A 2010 Sea-level rise and its impact on coastal zones *Science* **328** 1517-20

[8] Shahruzaila bin Abdullah 1993 Coastal developments in Malaysia: scope, issues and challenges. Department of Irrigation and Drainage Malaysia

[9] National Hydraulic Research Institute Malaysia 2010 The study of the Impact of Climate Change on Sea Level Rise on Malaysia Coastlines (Final Report) p172

[10] IPCC 2007 *Climate change 2007: the physical science basis. Contribution of working group I to the fourth assessment report of the IPCC*, ed S Solomon *et al* (Cambridge: Cambridge University Press)

[11] Department of Statistics Malaysia 2016 GDP by State - National Accounts - 2010-2015 p10

[12] Leete R 2011 Selangor's human development progress and challenges: UN Development Program

[13] Khairul Nizam Abdul Maulud and Siti Norsakinah Selamat 2018 Impact on the coastal development due to the shoreline changes *National seminar on Coastal Morphology*

[14] Khairul Nizam Abdul Maulud and Roshana Mar Rafar 2015 Determination the impact of sea level rise to shoreline changes using GIS *Int. Conf. on Space Science and Communication*

[15] Ho J C 2009 *Coastal floor risk assessment and coastal zone management: case study of Seberang Perai and Kuantan Pecan in Malaysia* Msc Thesis University of Southampton

[16] Ministry of Natural Resources and Environment Malaysia 2007 Flood and Drought Management in Malaysia. A Speech Draft of the Director of Ministry of Natural Resources and Environment Malaysia.
http://www.met.gov.my/files/ClimateChange2007/session1b/K2%20Husaini_p.doc

[17] Ahmad Shakir Mohd Saudi, Hafizan Juahir, Azman Azid and Fazureen Azaman 2015 Flood risk index assessment in Johor River Basin *Malaysian Journal of Analytical Sciences* **19 (5)** 991-1000

[18] Atikah Shafie, 2009 Technical report: extreme flood event, a case study on floods of 2006 and 2007 in Johor, Malaysia Msc thesis Colorado State University

[19] Baharuddin M K 2007 Climate Change: its effects on the agricultural sector in Malaysia *National Seminar on Socio-Economic Impacts of Extreme Weather and Climate Change*, 21-22 June, 2007 Malaysia

[20] Berita Harian 2007 Banjir: Kerajaan rugi RM1.5b. http://archives.emedia.com.my/bin/main.exe?f=doc&state=ltsa2h.2.25

[21] Cetdem 2010 *Socio-Economic Impacts of Climate Change in Malaysia* Center for Environment, Technology and Development Malaysia.

[22] Che Gayah Ismail 2011. Climate Services in Malaysia *Regional Seminar on Climate Services:* Malaysian Meteorological Department

[23] MERCY Malaysia Annual Report 2011. https://reliefweb.int/disaster/fl-2011-000021-mys/thumb

[24] Ayat K AB Rahman, Ramli Abdullah and Faizah Mohd Shariff 2012 The Economic impact of the north east monsoon and la nina on oil palm production in Malaysia *Oil Palm Industry Economic Journal* **12 (2)** 22-35

[25] Department of Irrigation and Drainage Malaysia 2007 Fenomena Banjir di Malaysia (Flood Phenomenon in Malaysia) http://www.water.gov.my/images/pdf/fenomenabanjir_msia.pdf

[26] Handmer J, Honda Y, Kundzewicz Z W, Arnell N, Benito G, Hatfield J, Mohamed I F, Peduzzi

P, Wu S, Sherstyukov B, Takahashi K and Yan Z 2012 *Changes in impacts of climate extremes: human systems and ecosystems in Managing the risks of extreme events and disasters to advance climate change adaptation.* ed C B Field et al. (Cambridge: Cambridge University Press) p 231–90

[27] Nicholls R J and Mimura N 1998 Regional issues raised by sea-level rise and their policy implications *Climate Research* **11** 5-18

[28] Ministry of Science, Technology and the Environment 2000 Malaysia Initial National Communication Submitted to the United Nations Framework Convention on Climate Change. http://unfccc.int/resource/docs/natc/malnc1.pdf

[29] Department of Irrigation and Drainage, Malaysia 2007 *Final report of national coastal vulnerability index study – phase 1.* Bureau for Innovation and Consultancy of Universiti Teknologi Malaysia for the Department of Irrigation and Drainage, Malaysia. Accessed from the repository of National Hydraulic Research Institute of Malaysia.

[30] Nor Aslinda bt Awang, Wan Hasliza bt Wan Jusoh and Yannie Anan Benson 2013 Impact of sea level rise on selected locations in Malaysia *Seminar on research in sea level rise and adaptation measures*

Nor Aslinda Awang and Elyasabath Mohd Ghazali 2013 Monitoring and evaluation of climate change adaptation initiatives for Malaysia *Apan-ICLEI- Beyond the Climate Change Adaptation Cycle*

[31] Yohe G W and Schesinger M E 1998 Sea-level rise change: the expected economic cost of protection or abandonment in the United States *Climatic Change* **38** 447-72

[32] Saizar A 1997 Assessment of a potential sea-level rise on the coast of Montevideo, Uruguay *Climate Research* **9(1-2)** 73-79

[33] Zeider R B 1997 Climate change vulnerability and response stratifies for the coastal zone of Poland. *Climatic Change* **36** 151-73

[34] Klein R J T and Tol R S J 1997 *Adaptation to climate change: options and technologies. An Overview Paper* Technical Paper FCCC/TP/1997/3 (Bonn, Germany: United Nations Framework Convention on Climate Change Secretariat) p36

[35] Tol R S J, Klein R J T, Jansen H M A and Verbruggen H 1996 Some economic considerations on the importance of proactive integrated coastal zone management *Ocean & Coastal Management* **32(1)** 39–55

[36] Linham M M and Nicholls R J 2010 Technologies for climate change adaptation – coastal erosion and flooding UNEP Risø Centre on Energy, Climate and Sustainable Development Risø DTU National Laboratory for Sustainable Energy, Roskilde Denmark

[37] Robert W S 2011 The scoping assessment on climate change adaptation in Malaysia: sustainable oceans and forests through equitable development (sofe dev) adaptation knowledge platform Bangkok, Thailand

[38] Ministry of Natural Resources and Environment 2010 Malaysia's Second National Communication (NC2) submitted to the United Nations Framework Convention on Climate Change (UNFCCC) http://nc2.nre.gov.my

IOP Conf. Series: Earth and Environmental Science **228** (2019) 012024 doi:10.1088/1755-1315/228/1/012024

So near yet so different: Surface ozone at three sites in Malaysia

F Ahamad[1], M T Latif[2], M F Yusoff[2], M F Khan[1], L Juneng[2]

[1] Centre for Tropical Climate Change System (IKLIM), Institute for Climate Change, Universiti Kebangsaan Malaysia, 43600 Bangi, Selangor, Malaysia
[2] School of Environmental and Natural Resource Sciences, Faculty of Science and Technology, Universiti Kebangsaan Malaysia, 43600 Bangi, Selangor, Malaysia

fatimah.a@ukm.edu.my

Abstract. Surface ozone is an air pollutant and greenhouse gas. This study attempts to identify some key challenges in ozone mitigation based on surface ozone observations at three locations within and near Klang Valley, Malaysia. Surface ozone data obtained from the Department of Environment Malaysia for Petaling Jaya (S1), Shah Alam (S2) and Cheras (S3) stations are analysed to show variation in ozone between stations that are in close proximity. Initial statistical analysis results indicate a large difference in the frequency of total daily maximum ozone concentration at S2 compared to S1 and S3 stations. The application of IoT could likely be a viable option for determining sensitivity to precursors and meteorological conditions once the technology has further matured.

1. Introduction

Surface ozone concentrations have been shown to be relatively high at certain locations in Malaysia when compared against the Malaysian Ambient Air Quality Guideline (MAAQG) of 100 pbb and 60 pbb for 1h and 8h ozone [1-3]. These high concentrations are a cause for concern as ozone has been shown to be detrimental to health as well as reduce crop yields [4, 5]. Mitigating this pollutant is a challenge as it involves complex interaction between precursors that form ozone and scavengers that reduce ozone concentrations as well as meteorological parameters that influence transport and photochemistry [6-8]. The concentration of ozone in many of the urban, suburban and industrial stations in Malaysia have been studied, with the focus mainly falling on the status of air quality, its sources or forecasting applications [9-11]. Given that ozone is a secondary pollutant which requires the presence of sunlight for its formation, identifying sources and making reliable forecasts presents unique challenges. In this study, surface ozone from observations from three monitoring stations located within a 20 km radius from each other is presented to highlight the challenges involved in ozone mitigation.

Content from this work may be used under the terms of the Creative Commons Attribution 3.0 licence. Any further distribution of this work must maintain attribution to the author(s) and the title of the work, journal citation and DOI.
Published under licence by IOP Publishing Ltd

2. Study area and methodology

The data source is the Department of Environment Malaysia. Three monitoring stations located within Peninsular Malaysia were selected for analysis due to their relative proximity to each other (Figure 1). These stations are located within the west coast of Peninsular Malaysia. S1 and S3 is located within a 10 km radius of S2. Ozone measurements from 2012 to 2014 were measured using Teledyne Model 400A, USA. The wind speed and direction for the period between June to August (JJA) 2014 were measured using Met One 010C sensor (wind speed), Met One 020C sensor (wind direction). The wind rose was plotted using WRPLOT View software Version 8.0.2 (https://www.weblakes.com) which was imported into Google Earth Pro Version 7.3.2.

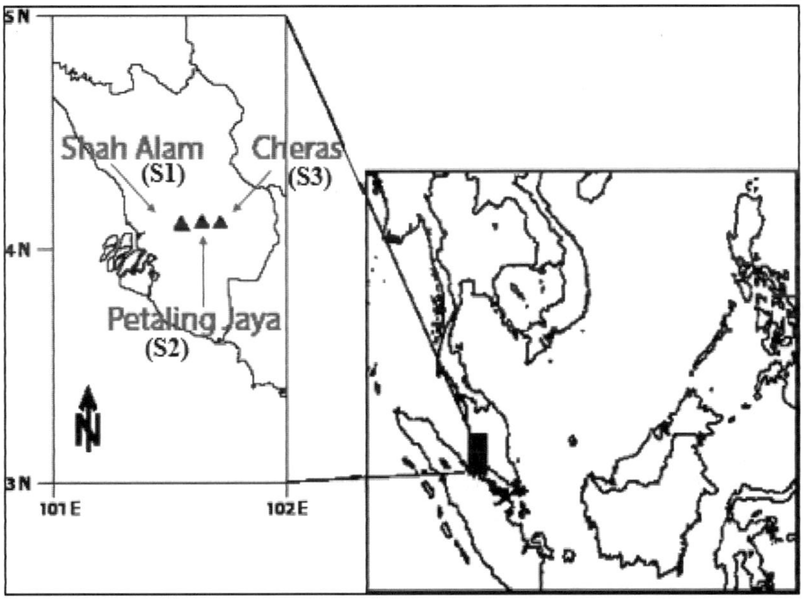

Figure 1. Location of monitoring stations

3. Results and discussion

The Malaysian Ambient Air Quality Guideline for 1h ozone is 100 ppb. The number of times ozone concentration equaled or exceeded this value between 2012 and 2014 is presented in Figure 2. The results indicate much fewer exceedance at S2 compared to S1 and S3. Similar pattern has been observed in short and long-term observations at these locations [1, 3]. The pattern of ozone exceedance between the years also differ between all three monitoring stations although the distance between the stations are less than 10 km apart. One of the likely reasons proposed for the very low exceedance as S2 is its location near a busy intersection that is conducive to ozone titration [1]. However, titration alone does not help explain the difference in ozone exceedance pattern between the years.

IOP Conf. Series: Earth and Environmental Science **228** (2019) 012024 doi:10.1088/1755-1315/228/1/012024

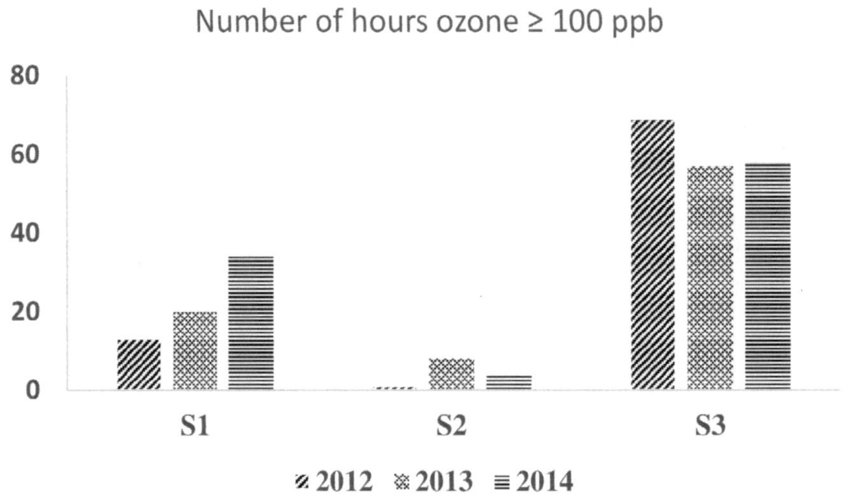

Figure 2. The number of times ozone ≥ 100 ppb at the monitoring stations between 2012 and 2014

To identify the likely influence of meteorology, Figures 3(a) and 3(b) were plotted to compare wind profiles at S2 and S3 during the south west monsoon season. Although the prevalent wind conditions should be the south west direction, near surface measurements (10 m) indicated large difference in the wind profile. Not only is the prevalent wind direction different the period of calm winds was very different as well (2% at S2 and 40% at S3). The land use type (historical) also differ with a much more compact structure and (closer) to major roads at S2 compared to S3.

Figure 3(a). Wind rose profile at S2 for the period JJA (2014) and the satellite image at location.

Figure 3(b). Wind rose profile at S3 for the period JJA (2014) and the satellite image at location

Although some of these observation pattern has been highlighted in previous studies, there has not been focused research on the mitigation implication of these observations in Malaysia. If titration of ozone is the predominant factor determining ozone concentration in S2, control strategies involving NOx control would increase ozone in S2. The matter is further complicated by the fact that based on political boundary, S1 and S2 fall under a different local jurisdiction from S3 although S1 and S2 have a closer ozone profile. Even if a blanket policy such as ban on vehicle entry is applied, the effect on ozone is unclear without information on its precursor sensitivity (NOx sensitive or VOC sensitive where NOx represents oxides of nitrogen and VOC represent Volatile Organic Compounds).

4. Way forward

Historical database on ozone and NOx distribution is insufficient to answer some of the questions needed to identify potential effects of specific NOx and VOC control strategies. Further research on ozone precursor (NOx and VOC) control strategies as well as the specific meteorological contribution would be necessary for identifying the best strategy for overall ozone reduction. The challenges in ozone mitigation could also be addressed with the application of Internet of Things whereby citizen scientist function as data providers to provide higher spatial resolution to available measurements and/or parameters previously not measured. Ideally low-cost sensors can be deployed within a grid and the real time measurements be used not only to forecast potential ozone episodes but also provide sufficient historical data to identify best control strategies and monitor the relative success of these strategies.

References

[1] Ahamad F, Latif M T, Tang R, Juneng L, Dominick D, and Juahir H, 2014, *Variation of surface ozone exceedance around Klang Valley, Malaysia*. Atmospheric Research. **139**, 116-127.

[2] Latif M T, Huey L S, and Juneng L, 2012, *Variations of surface ozone concentration across the Klang Valley, Malaysia*. Atmospheric Environment. **61**(0), 434-445.

[3] Mohtar A A A, Latif M T, Baharudin N H, Ahamad F, Chung J X, Othman M, and Juneng L, 2018, *Variation of major air pollutants in different seasonal conditions in an urban environment in Malaysia*. Geoscience Letters. **5**(1), 21.

[4] Hanna A F, et al., 2011, *Associations between ozone and morbidity using the Spatial Synoptic Classification system*. Environmental Health. **10**(1), 49.

[5] Ishii S, Bell J, and Marshall F, 2007, *Phytotoxic risk assessment of ambient air pollution on agricultural crops in Selangor State, Malaysia*. Environmental Pollution. **150**(2), 267-279.

[6] Monks P S, et al., 2009, *Atmospheric composition change – global and regional air quality*. Atmospheric Environment. **43**(33), 5268-5350.

[7] Sillman S, 1999, *The relation between ozone, NOx and hydrocarbons in urban and polluted rural environments*. Atmospheric Environment. **33**, 1821-1845.

[8] Pusede S E, Steiner A L, and Cohen R C, 2015, *Temperature and recent trends in the chemistry of continental surface ozone*. Chemical reviews. **115**(10), 3898-3918.

[9] Azmi S Z, Latif M T, Ismail A S, Juneng L, and Jemain A A, 2010, *Trend and status of air quality at three different monitoring stations in the Klang Valley, Malaysia*. Air Quality, Atmosphere & Health. **3**, 53-64.

[10] Ismail M, Ibrahim M Z, Ibrahim T A, and Abdullah A M, 2011, *Time Series Analysis of Surface Ozone Monitoring Records in Kemaman, Malaysia*. Sains Malaysiana. **40**(5), 411-417.

[11] Awang N R, Elbayoumi M, Ramli N A, and Yahaya A S, 2016, *The influence of spatial variability of critical conversion point (CCP) in production of ground level ozone in the context of tropical climate*. Aerosol and Air Quality Research. **16**(1), 153-165.

Acknowledgements

The authors would like to thank the Department of Environment Malaysia for the provision of data. The authors also would like to acknowledge Geran Galakan Penyelidik Muda GGPM-2017-083.

Institute of Physics
Dirac House, Temple Back
Bristol BS1 6BE UK

ISSN: 1755-1307
ISBN 978-1-5108-8156-3

CURRAN ASSOCIATES INC.
proceedings.com